Student Solutions Manual

An Introduction to Statistical Methods and Data Analysis

SEVENTH EDITION

R. Lyman Ott

Michael Longnecker
Texas A&M University

Prepared by

John Daniel Draper
The Ohio State University

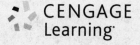

Australia • Brazil • Mexico • Singapore • United Kingdom • United States

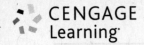

© 2016 Cengage Learning

WCN: 01-100-101

ALL RIGHTS RESERVED. No part of this work covered by the copyright herein may be reproduced, transmitted, stored, or used in any form or by any means graphic, electronic, or mechanical, including but not limited to photocopying, recording, scanning, digitizing, taping, Web distribution, information networks, or information storage and retrieval systems, except as permitted under Section 107 or 108 of the 1976 United States Copyright Act, without the prior written permission of the publisher.

For product information and technology assistance, contact us at
**Cengage Learning Customer & Sales Support,
1-800-354-9706**.

For permission to use material from this text or product, submit all requests online at **www.cengage.com/permissions**
Further permissions questions can be emailed to
permissionrequest@cengage.com.

ISBN: 978-1-305-26948-4

Cengage Learning
20 Channel Center Street
Boston, MA 02210
USA

Cengage Learning is a leading provider of customized learning solutions with office locations around the globe, including Singapore, the United Kingdom, Australia, Mexico, Brazil, and Japan. Locate your local office at: **www.cengage.com/global**.

Cengage Learning products are represented in Canada by Nelson Education, Ltd.

To learn more about Cengage Learning Solutions, visit **www.cengage.com**.

Purchase any of our products at your local college store or at our preferred online store **www.cengagebrain.com**.

Printed in the United States of America
Print Number: 01   Print Year: 2015

Contents

Chapter 1: What is Statistics 1

Chapter 2: Using Surveys and Experimental Studies to Gather Data 2

Chapter 3: Data Description 7

Chapter 4: Probability and Probability Distributions 25

Chapter 5: Inferences about Population Central Values 35

Chapter 6: Inferences Comparing Two Population Central Values 44

Chapter 7: Inferences about Population Variances 56

Chapter 8: Inferences about More Than Two Population Central Values 63

Chapter 9: Multiple Comparisons 77

Chapter 10: Categorical Data 83

Chapter 11: Linear Regression and Correlation 102

Chapter 12: Multiple Regression and the General Linear Model 122

Chapter 13: Further Regression Topics 152

Chapter 14: Analysis of Variance for Completely Randomized Designs 187

Chapter 15: Analysis of Variance for Blocked Designs 201

Chapter 16: The Analysis of Covariance 212

Chapter 17: Analysis of Variance for Some Fixed-, Random-, and Mixed-Effects Models 221

Chapter 18: Split-Plot, Repeated Measures, and Crossover Designs 235

Chapter 19: Analysis of Variance for Some Unbalanced Designs 249

Chapter 1

Statistics and the Scientific Method

1.1
- a. The population of interest is all salmon released from fish farms located in Norway.
- b. The samples are the two batches of salmon released (1,996 and 2,499 in northern and southern Norway, respectively).
- c. The migration pattern and survival of salmon released from fish farms.
- d. Since the sample is only a small proportion of the whole population, it is necessary to evaluate what the mean weight may be for any other random selection of farmed salmon.

1.3
- a. All families that have had option of SNAP (food stamps).
- b. 60,782 examined over the time period of 1968 to 2009.
- c. Adult health and economic outcomes (specifically, the incidence of metabolic health outcomes and economic self-sufficiency).
- d. In order to evaluate how closely the sample families represent the American population over this time period.

1.5
- a. The population of interest is the population of those who would vote in the 2004 senatorial campaign.
- b. The population from which the sample was selected is registered voters in this state.
- c. The sample will adequately represent the population, unless there is a difference between registered voters in the state and those who would vote in the 2004 senatorial campaign.
- d. The results from a second random sample of 5,000 registered voters will not be exactly the same as the results from the initial sample. Results vary from sample to sample. With either sample we hope that the results will be close to that of the views of the population of interest.

Chapter 2

Using Surveys and Experimental Studies to Gather Data

2.1
- a. The explanatory variable is level of alcohol drinking. One possible confounding variable is smoking. Perhaps those who drink more often also tend to smoke more, which would impact incidence of lung cancer. To eliminate the effect of smoking, we could block the experiment into groups (e.g., nonsmokers, light smokers, heavy smokers).
- b. The explanatory variable is obesity. Two confounding variables are hypertension and diabetes. Both hypertension and diabetes contribute to coronary problems. To eliminate the effect of these two confounding variables, we could block the experiment into four groups (e.g., hypertension and diabetes, hypertension but no diabetes, diabetes but no hypertension, neither hypertension nor diabetes).

2.3 Possible confounding factors include student-teacher ratios, expenditures per pupil, previous mathematics preparation, and access to technology in the inner city schools. Adding advanced mathematics courses to inner city schools will not solve the discrepancy between minority students and white students, since there are other factors at work.

2.5 The relative merits of the different types of sampling units depends on the availability of a sampling frame for individuals, the desired precision of the estimates from the sample to the population, and the budgetary and time constraints of the project.

2.7
- a. All residents in the county.
- b. All registered voters.
- c. Survey nonresponse – those who responded were probably the people with much stronger opinions than those who did not respond, which then makes the responses not representative of the responses of the entire population.

2.9
- a. Alumni (men only?) who graduated from Yale in 1924.
- b. No. Alumni whose addresses were on file 25 years later would not necessarily be representative of their class.
- c. Alumni who <u>responded</u> to the mail survey would not necessarily be representative of those who were <u>sent</u> the questionnaires. Income figures may not be reported accurately (intentionally), or may be rounded off to the nearest $5,000, say, in a self-administered questionnaire.
- d. Rounding income responses would make the figure $25,111 unlikely. The fact that higher income respondents would be more likely to respond (bragging), and the fact that incomes are likely to be exaggerated, would tend to make the estimate too high.

2.11
   a. Simple random sampling.
   b. Stratified sampling.
   c. Cluster sampling.

2.13
   a. Stratified sampling. We should stratify by type of degree and then sample 5% of the alumni within each degree type. This method will allow us to examine the employment status for each degree type and compare among them.
   b. Simple random sampling. Once we find 100 containers we will stop. Still it will be difficult to get a completely random sample. However, since we don't know the locations of the containers, it would be difficult to use either a stratified or cluster sample.

2.15 This is an example where there are two levels of Experimental units. The analysis is discussed in Chapter 18.

   To study the effect of month:
   a. Factors: Month
   b. Factor levels: 8 levels of month (Oct - May)
   c. Block = each section
   d. Experimental unit (Whole plot EU) = each tree
   e. Measurement unit = each orange
   f. Replications = 8 replications of each month
   g. Covariates = none
   h. Treatments = 8 treatments (Oct – May)

   To study the effect of location:
   a. Factors: Location
   b. Factor levels: 3 levels of location (top, middle, bottom)
   c. Block = each section
   d. Experimental unit (Split plot EU) = each location tree
   e. Measurement unit = each orange
   f. Replications = 8 replications of each location
   g. Covariates = none
   h. Treatments = 3 treatments (top, middle, bottom)

2.17
   a. Factors: Type of treatment
   b. Factor levels: D_1, D_2, Placebo
   c. Blocks: Hospitals, Wards
   d. Experimental units: Patients
   e. Measurement units: Patients
   f. Replications: 2 patients per drug in each of the ward/hospital combinations
   g. Covariates: None
   h. Treatments: D_1, D_2, Placebo

2.19
- a. Factors: Temperature, Type of seafood
- b. Factor levels: Temperature (0 °C, 5 °C, 10 °C); Type of seafood (oysters, mussels)
- c. Blocks: None
- d. Experimental units: Package of seafood
- e. Measurement units: Sample from package
- f. Replications: 3 packages per temperature
- g. Treatments: (0 °C, oysters), (5 °C, oysters), (10 °C, oysters), (0 °C, mussels), (5 °C, mussels), (10 °C, mussels)

2.21
- a. Randomized complete block design with blocking variable (10 warehouses) and 5 treatments (5 vendors)

2.23
- a. Design B. The experimental units are not homogeneous since one group of consumers gives uniformly low scores and another group gives uniformly high scores, no matter what recipe is used. Using design A, it is possible to have a group of consumers that gives mostly low scores randomly assigned to a particular recipe. This would bias this particular recipe. Using design B, the experimental error would be reduced since each consumer would evaluate each recipe. That is, each consumer is a block and each of the treatments (recipes) is observed in each block. This results in having each recipe subjected to consumers who give low scores and to consumers who give high scores.
- b. This would not be a problem for either design. In design A, each of the remaining 4 recipes would still be observed by 20 consumers. In design B, each consumer would still evaluate each of the 4 remaining recipes.

2.25
- a. Each state agency and some federal agencies have records of licensed physicians, professional corporations, facility licenses, etc. Professional organizations such as the American Medical Association, American Hospital Administrators Association, etc., may have such lists, but they may not be as complete as licensing records.
- b. What nursing specialties are available at this time at the physician's offices or medical facilities? What medical specialties/facilities do they anticipate adding or expanding? What staffing requirements are unfilled at this time or may become available when expansion occurs? What is the growth/expansion time frame?
- c. Licensing boards may have this information. Many professional organizations have special categories for members who are unemployed, retired, working in fields not directly related to nursing, students who are continuing their education, etc.
- d. Population growth estimates may be available from the Census Bureau, university economic growth research, bank research studies (prevailing and anticipated load patterns), etc. Health risk factors and location information would be available from state health departments, the EPA, epidemiological studies, etc.
- e. Licensing information should be stratified by facility type, size, physician's specialty, etc., prior to sampling.

2.27

| | Factor 2 | | |
Factor 1	I	II	III
A	25	45	65
B	10	30	50

2.29
   a. Bake one cake from each recipe in the oven at the same time. Repeat this procedure r times. The baking period is a block with the four treatments (recipes) appearing once in each block. The four recipes should be randomly assigned to the four positions, one cake per position. Repeat this procedure r times.
   b. If position in the oven is important, then position in the oven is a second blocking factor along with the baking period. Thus, we have a Latin square design. To have $r = 4$, we would need to have each recipe appear in each position exactly once within each of four baking periods. For example:

Period 1	Period 2	Period 3	Period 4
R_1 R_2	R_4 R_1	R_3 R_4	R_2 R_3
R_3 R_4	R_2 R_3	R_1 R_2	R_4 R_1

   c. We now have an incompleteness in the blocking variable period since only four of the five recipes can be observed in each period. In order to achieve some level of balance in the design, we need to select enough periods in order that each recipe appears the same number of times in each period and the same total number of times in the complete experiment. For example, suppose we wanted to observe each recipe $r = 4$ times in the experiment. If would be necessary to have 5 periods in order to observe each recipe 4 times in each of the 4 positions with exactly 4 recipes observed in each of the 5 periods.

Period 1	Period 2	Period 3	Period 4	Period 5
R_1 R_2	R_5 R_1	R_4 R_5	R_3 R_4	R_2 R_3
R_3 R_4	R_2 R_3	R_1 R_2	R_5 R_1	R_4 R_5

2.31
   a. All cars (and by extension, their tires) in the state.
   b. Cars registered in the 4 months in which the sample was taken.
   c. 2 potential concerns arise: not all cars in the region are registered and the time of year may lead to ignoring some cars (some people leave the area for the winter). Unregistered cars may have a higher proportion of unsafe tire tread thickness.

2.33
   a. People are notoriously bad at recall. A telephone interview immediately following the time of interest would likely be best, but nonresponse is often high. Mailed questionnaires would likely be administered too late to be of use and personal interviewing would be intractable to interview in a timely manner.
   b. All three are potential avenues. Interviews are more personal but more time consuming. Mailing questionnaires should also work as the editor has a list of his/her clientele, but if he wants to garner information about perspectives of those not reading his/her paper, he/she may need to blanket the city with questionnaires. Telephone interviews may be difficult as finding the numbers of those in the area may be difficult.

6 *Chapter 2: Using Surveys and Experimental Studies to Gather Data*

   c. Again, all three methods would be viable. A mailed questionnaire would be the easiest and cheapest but the response rate would likely be lower.
   d. If the county believes they have an accurate list of those with dogs, a mailed questionnaire or telephone interview would work, but using a list of registered dogs may be underrepresenting those who haven't taken good care of their dogs (and thereby underrepresenting the proportion with rabies shots).

Chapter 3

Data Description

3.1
  a. The following is a pie chart of the federal expenditures for the 2014 fiscal year (in billions of dollars).

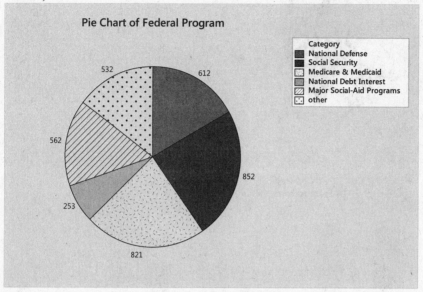

  b. The following is a bar chart of the federal expenditures for the 2006 fiscal year (in billions of dollars).

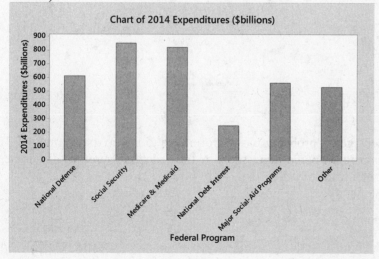

c. The following are a pie chart and bar chart of the federal expenditures for the 2014 fiscal year (in percentages).

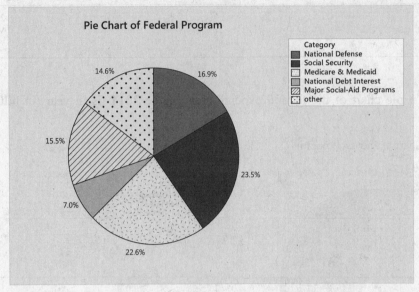

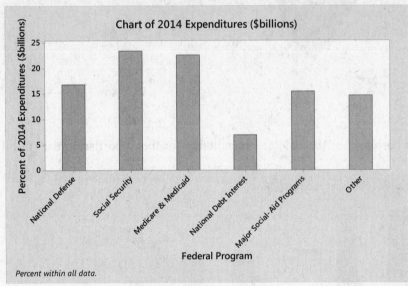

d. The pie chart using percentages is probably most informative to the tax-paying public. Here the tax-paying public can compare the percentages spent by the Federal government for domestic and defense programs as part of a whole.

3.3
a. The following bar chart shows the increase in the number of family practice physicians (in thousands of physicians) over the period 1980-2001.

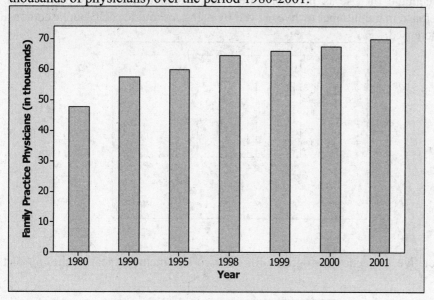

b. The percent of office-based physicians who are family practice physicians over the period 1980-2001 can be seen in the following table.

	1980	1990	1995	1998	1999	2000	2001
Percent Family Practice	17.6	16.0	14.0	13.8	14.0	13.8	13.6

The following bar chart shows the percent of office-based physicians who are family practice physicians over the period 1980-2001.

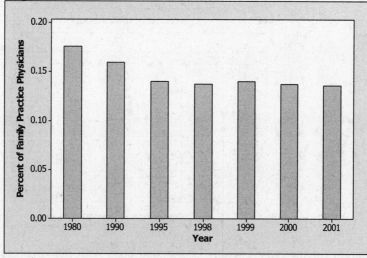

c. While the number of family practice physicians increased over the period 1980-2001, the percent of total office-based physicians who are family practice physicians decreased over the same period.

10   Chapter 3: Data Description

3.5 Two separate bar graphs could be plotted, one with Lap Belt Only and the other with Lap and Shoulder Belt. A single bar graph with the Lap Belt Only value plotted next to the Lap and Shoulder Belt for each value of Percentage of Use is probably the most effective plot. This plot would clearly demonstrate that the increase in the number of lives saved by using a shoulder belt increased considerably as the percentage use increased.

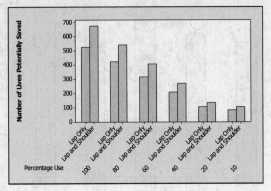

3.7
    a. The separate relative frequency histograms for the two treatments appear at the top of the next page.

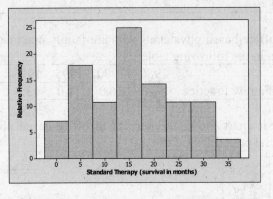

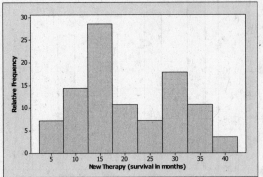

    b. The histogram for the New Therapy begins and ends with bins that are slightly higher than the bins in the histogram for the Standard Therapy. This would indicate that the New Therapy generates a few more large values than the Standard Therapy. However, there is not convincing evidence that the New Therapy generates a longer survival time.

3.9
a. The time series plot shows an increase from 1980 through 1990, with a large dip at 1991 (after the Persian Gulf Conflict). There is then a decrease in expenditures in billions of dollars over the period 1992 to 1999, and then a sharp increase from 1999 through 2012.

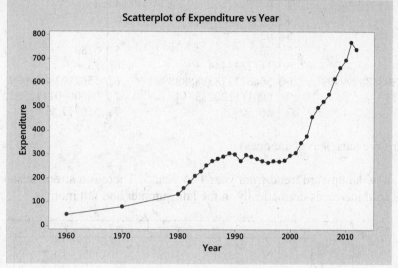

b. The time series plot of expenditures as a percent of GNP shows a large decrease from 1960-1980 and an increase from 1980 to 1983. It is then fairly steady from 1983 to 1987, decreases from 1987 to 2001 (with the exception of a spike in 1992), has a sharp increase 2001 to 2002, and then is fairly steadily increasing from 2002 to 2012. The plot appears on the next page.

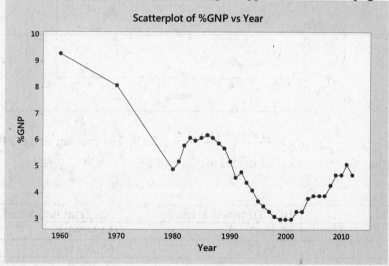

c. The plots do not show similar trends. The time series plot of expenditures supports the assertions of these members of Congress, but the %GDP vs. Year plot appears to contradict those liberal members.
d. The expenditures show a steady rise due to inflation. The %GDP plot is much more telling of the landscape. % military spending show increases that can be related to various historical events. The increase in the early 80s is likely due to the Iran Contra affair. 91-92 shows a spike for Operation Desert Storm. The recent increasing trend is likely explained by the War on Terror.

3.11 The following are stem-and-leaf plots for each of the three years 1985, 1996, and 2002.

	1985		1996		2002
3	7	3		3	
4		4	0	4	4
4		4		4	
5	014	5	02	5	
5	7	5	56	5	5789
6	00011122334	6	1112233444	6	23
6	5566677778888889999999	6	56667777888888888999	6	55677778899999
7	000000111233	7	000111122233344	7	00000000111222223333334444
7	5	7	56	7	556777

3|7 = 37% (stems are tens, leaves are ones)

3.13 The plots show an upward trend from year 1 to year 5. There is a strong seasonal (cyclic) effect; the number of units sold increases dramatically in the late summer and fall months.

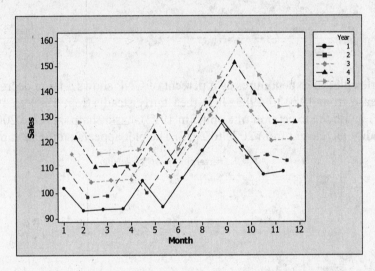

3.15 The mean is $\bar{y} = \frac{1}{n}\sum_{i=1}^{21} y_i = \frac{35+81+96+\cdots+24}{21} = 55.19$. The median is the 11th value when arranged in increasing order: median = 58. The modes are 24 and 58.

3.17

	10% Trimmed Mean	5% Trimmed Mean
3.15	53.5647	54.9789
3.16	53.5647	68.2421

The 10% trimmed mean for 3.15 and 3.16 are the same because the outliers are both trimmed off. The 5% trimmed mean differs as only the highest and lowest value are trimmed so one of the outliers is still accounted for in the 5% trimmed mean.

3.19
a. The relative frequency histogram is unimodal and slightly right skewed.

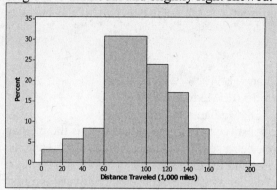

b. The following table is used to calculate the summary statistics:

Class Interval	Frequency (f_i)	Midpoint (y_i)	$f_i y_i$
0-20.0	6	10	60
20.1-40.0	11	30	330
40.1-60.0	16	50	800
60.1-100.0	59	80	4720
100.1-120.0	46	110	5060
120.1-140.0	33	130	4290
140.1-160.0	16	150	2400
160.1-200.0	4	180	720
Total	191		18,380

mean $\approx 18380/191 = 96.2$
mode ≈ 80

$$\text{median} \approx L + \frac{w}{f_m}(0.5n - cf_b) = 100.1 + \frac{20}{46}[(0.5)(191) - 92] = 101.6$$

c. Since the median is larger than the mean, it would indicate that the plot is somewhat left skewed. This contradiction between what is indicated in the relative frequency histogram and what is indicated by the summary statistics is due to the fact that the class intervals are a different width. The correct plot would have relative frequency divided by class width on the vertical axis. This would then produce a left skewed histogram with mode at approximately 110.

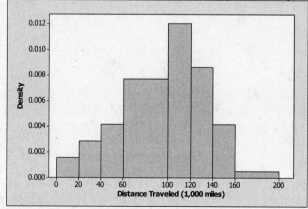

14 Chapter 3: Data Description

d. The median is more informative since the distribution is somewhat skewed to the left which produces a mean somewhat less than the middle of the distribution. The median distance traveled would at least represent a value such that half of the buses traveled less and half greater than 101,600 miles.

3.21
a. Mean = 8.04, Median = 1.54
b. Terrestrial: Mean = 15.01, Median = 6.03
   Aquatic: Mean = 0.38, Median = 0.375
c. The mean is more sensitive to extreme values than is the median.
d. Terrestrial: Median, because the two large values (76.50 and 41.70) result in a mean that is larger than 82% of the values in the data set.
   Aquatic: Mean or median since the data set is relatively symmetric.

3.23
a. If we use all 14 failure times, we obtain Mean > 173.7 days and Median = 154 days. In fact, we know that the mean is greater than 173.7 days since the failure times for two of the engines are greater than the reported times of 300 days.
b. The median would be unchanged if we replace the failure times of 300 with the true failure times for the two engines that did not fail. However, the mean would be increased.

3.25 Mean = 1.7707, Median = 1.7083, Mode = 1.273
The average of the three net group means and the mean of the complete set of measurements are the same. This will be true whenever the groups have the same number of measurements, but it is not true if the groups have different sample sizes. However, the average of the group medians and modes are different from the overall median and mode.

3.27
a. $s = 7.95$ years
b. Because the magnitude of the racers' ages is larger than that of their experience.

3.29 The quantile plot is given below.

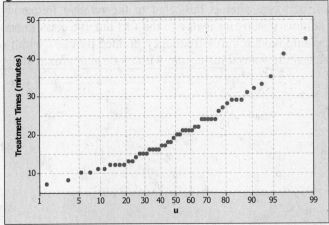

a. The 25th percentile is the value associated with $u = 0.25$ on the graph, which is 14 minutes. Also, by definition 14 minutes is the 25th percentile since 25% of the times are less than or equal to 14 minute and 75% of the times are greater than or equal to 14 minutes.

b. Yes; the 90th percentile is 31.5 minutes. This means that 90% of the patients have a treatment time less than or equal to 31.5 minutes (which is less than 40 minutes).

3.31
 a. Luxury: $\bar{y} = 145.0$, $s = 27.6$
    Budget: $\bar{y} = 46.1$, $s = 5.13$
 b. Luxury: CV = 19%
    Budget: CV = 11%
 c. Luxury hotels vary in quality, location, and price, whereas budget hotels are more competitive for the low-end market so prices tend to be similar.
 d. The CV would be better because it takes into account the larger difference in the means between the two types of hotels.

3.33

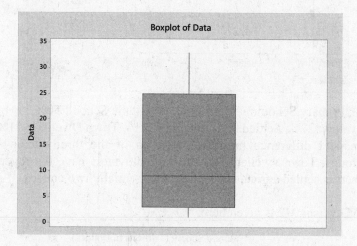

3.35
 a. CAN: $Q_1 \approx 1.45$, Q_2 = Median ≈ 1.65, $Q_3 \approx 2.4$
    DRY: $Q_1 \approx 0.55$, Q_2 = Median ≈ 0.60, $Q_3 \approx 0.70$
 b. Canned dog food is more expensive (median much greater than that for dry dog food), highly skewed to the right with a few large outliers. Dry dog food is slightly left skewed with a considerably less degree of variability than canned dog food.

3.37
 a. 1985: mean = 65.876; median = 67.90
    1996: mean = 66.843; median = 68.20
    2002: mean = 69.449; median = 70.20
    Since the distributions are left skewed, it is better to use the median for each of the three years.
 b. 1985: $s = 6.734$; CV = 10.2%
    1996: $s = 6.688$; CV = 10.0%
    2002: $s = 6.163$; CV = 8.9%
    The coefficients of variation are decreasing over the three years.

3.39
a. A stacked bar graph is given below.

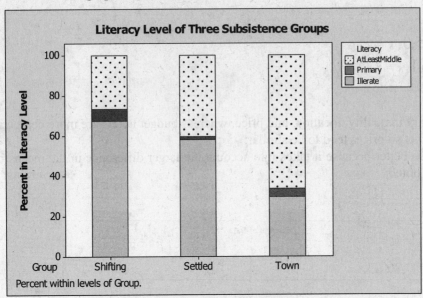

b. Illiterate: 46%; Primary Schooling: 4%; At Least Middle School: 50%
Shifting Cultivators: 27%; Settled Agriculturists: 21%; Town Dwellers: 51%
There is a marked difference in the distribution of the three literacy levels for the three subsistence groups. Town dwellers and shifting cultivators have the reverse trends in the three categories, whereas settled agriculturists fall into essentially two classes.

3.41 A scatterplot of M3 versus M2 is given here.

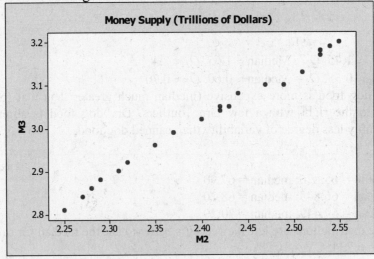

a. Yes, it would since we want to determine the relative changes in the two over the 20-month period of time.
b. See scatterplot. The two measures follow an increasing, approximately linear relationship.

3.43
   a. Mean = 57.5; Median = 34.0
   b. Median since the data has a few very large values which results in the mean being larger than all but a few of the data values.
   c. Range = 273; $s = 70.2$
   d. Using the approximation, $s \approx \text{range}/4 = 273/4 = 68.3$. The approximation is fairly accurate.
   e. $\bar{y} \pm s \Rightarrow (-12.7, 127.7)$; yields 82%
      $\bar{y} \pm 2s \Rightarrow (-82.9, 197.0)$; yields 94%
      $\bar{y} \pm 3s \Rightarrow (-153.1, 268.1)$; yields 97%
   f. The Empirical Rule applies to data sets with roughly a "mound-shaped" histogram. The distribution of this data set is highly skewed right.

3.45
   a. Price per roll: Mean = 0.9196, $s = 0.4233$
      Price per sheet: Mean = 0.01091, $s = 0.0059$
   b. Price per roll: $CV = 100 \dfrac{0.4233}{0.9196}\% = 46.03\%$

      Price per sheet: $CV = 100 \dfrac{0.0059}{0.01091}\% = 54.13\%$

      The price per sheet is more variable relative to its mean.
   c. CV; The CV is unit free, whereas the standard deviation also reflects the relative magnitude of the data values.

3.47 Boxplots for price per roll and number of sheets per roll are given here.

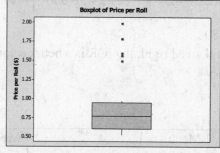

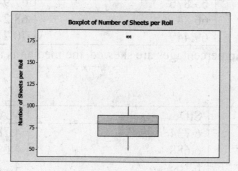

From the two boxplots, there a 5 unusual brands with regard to price per roll: $1.49, $1.56, $1.59, $1.78, and $1.98. There are 2 unusual brands with respect to number of sheets per roll: 180 and 180.

3.49 A relative frequency histogram for murder rate is given here.

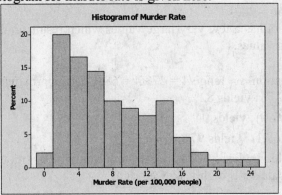

3.51

a. 75th percentile $\approx L + \dfrac{w}{f_m}\left[0.75n - cf_b\right] = 13.5 + \dfrac{2}{9}\left[0.75(90) - 72\right] = 12.5$

25th percentile $\approx L + \dfrac{w}{f_m}\left[0.25n - cf_b\right] = 3.5 + \dfrac{2}{15}\left[0.25(90) - 20\right] = 3.83$

IQR $\approx 12.5 - 3.83 = 8.67$

$s^2 \approx \dfrac{1}{n-1}\sum_{i=1}^{13} f_i(y_i - 8.3)^2$

$= \dfrac{1}{89}\left[2(0.5-8.3)^2 + 18(2.5-8.3)^2 + \cdots + 1(24.5-8.3)^2\right] = 29.0382$

Thus, $s \approx \sqrt{29.0382} = 5.389$.

b. Because the distribution is skewed right, the IQR is a better (more robust) measure of spread.
c. The population is all SMSAs.

3.53

a.

Year	Mean	Median
1985	65.876	67.9
1996	66.843	68.2
2002	69.449	70.2

b. Since the homeownership percentages are skewed, the median is a more reliable measure of center.

c.

Year	StDev	MAD
1985	6.734	4.783
1996	6.688	4.697
2002	6.163	4.229

d. Skewed data is better served by a robust measure like MAD.
e. It appears home ownership rates are increasing and the variability is (slightly

3.55
a. The means are given here:

Number of Members	Mean
1	93.75
2	98.652
3	113.3125
4	124.857
5+	131.90

b. Mean = 9024/83 = 108.7228
c. Yes, we can use the following method:
[20(93.75) + 23(98.652) + 16(113.3125) + 14(124.857) + 10(131.90)]/83
= 9023.994/83 = 108.7228
d. As the number of members increases, the mean expenditures tends to increase.

3.57
a. The sample mean will be distorted by several large values which skew the distribution. State 5 and State 11 have more than 10 times as many plants destroyed as any other state; for arrests, States 1, 2, 8, and 12 exceed the other arrest figures substantially.
b. Plants: $\bar{y} = 10,166,919/15 = 677,794.60$

Arrests: $\bar{y} = 1425/15 = 95$

10% trimmed mean:
Plants: $\bar{y} = 1,565,604/11 = 142,327.64$

Arrests: $\bar{y} = 657/11 = 59.7$

20% trimmed mean:
Plants: $\bar{y} = 1,197,354/9 = 133,039.33$

Arrests: $\bar{y} = 372/9 = 41.30$

For plants, the 10% trimmed mean works well since it eliminates the effect of States 5 and 11. For arrests, the means differ because each takes some of the high values out of the calculation. It appears that the distribution is not skewed, but rather separated into at least two parts: states with high numbers of arrests and states with low numbers. By trimming the mean, we may be losing important information.

3.59
a. The job-history percentages within each source are given here:

	Source		
Job History	Within Firm	Related Business	Unrelated Business
Promoted	22.80	19.05	23.81
Same position	56.14	38.10	42.86
Resigned	15.80	28.57	23.81
Dismissed	5.26	14.28	9.52
Total	100 ($n = 57$)	100 ($n = 21$)	100 ($n = 42$)

b. If for each source we compute the percentages combined over the promoted and same position categories, we find that they are 78.94% for within firms, 57.15% for related business, and 66.67% for unrelated business. This ordering by source also holds for every job history category except the "promoted" one in which the three sources are nearly equal. It appears that a company does best when it selects its middle managers from within its own firm and worst when it takes its choices from a related firm.

c. The stacked bar chart gives us a visual representation of how hiring source may (or may not) be related to job trajectory. It appears hiring from within leads to a more successful middle manager.

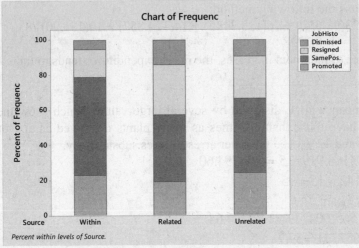

3.61 Arbitration seems to win the largest wage increases. If we assume that the Empirical Rule holds for these data, then a standard error for the mean of the arbitration figures would be $s/\sqrt{n} = 0.25$. Thus the mean increase after arbitration is $(9.42 - 8.40)/0.25 \approx 4$ standard errors above the next largest mean, that for poststrike. Management, on the other hand, should favor negotiation. It has the smallest mean percentage wage increase and the smallest variance, or least risk.

3.63
The treatment group has a stronger linear relationship with the baseline for each of the four clinic visits than does the placebo group, especially for the third clinic visit (Y_3). The placebo group has a weak positive relationship with age for all four clinic visits, while the treatment group has a weak negative relationship with age for all four clinic visits.

3.65 Scatterplots for math versus %poverty and %minority are given here:

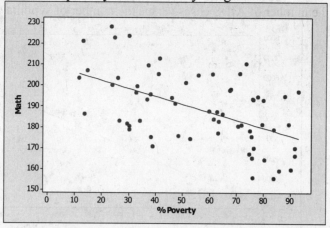

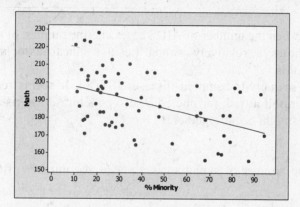

a. The scatterplots for math versus %poverty and %minority are very similar to the scatterplots for reading versus the same two variables (as shown on page 115 of the text). Thus, there is support for the same conclusions for the math scores as for the reading scores.
b. The conclusions are not different.

3.67 A time-series plot for calorie intake is given here:

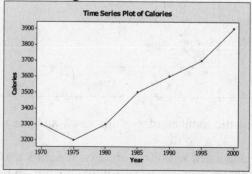

a. Aside from a slight dip at 1975, calorie intake has increased over the 30 years.
b. If we assume that calories would continue increasing through 2005, we would predict that the calorie intake would be about 4000 calories.

Chapter 3: Data Description

3.69
   a. A scatterplot of the number of AIDS cases versus the number of syphilis cases is given here:

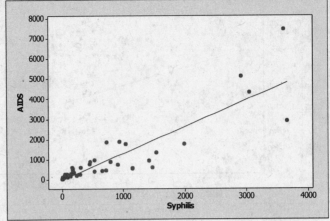

   b. The correlation between the number of AIDS cases and the number of syphilis cases is 0.883.
   c. The scatterplot shows a relatively strong positive linear association, and the correlation coefficient reflects this.
   d. Both diseases are sexually-transmitted diseases (STDs). It seems reasonable to believe that a person who puts himself at risk for one STD also puts himself at risk for another STD and that incidences of STDs might occur together.

3.71
   a. A scatterplot of the number of syphilis cases versus the number of tuberculosis cases is given here:

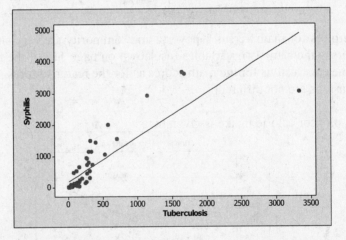

   b. The correlation between the number of syphilis cases and the number of tuberculosis cases is 0.849.
   c. Both diseases may be associated with drug abusers. It seems reasonable to believe that a person who puts himself at risk for one on disease through drug abuse also puts himself at risk for another disease and that incidences of such diseases might occur together.

3.73
a. A quantile plot of the number of tuberculosis cases is given here:

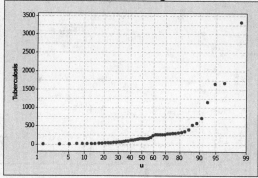

b. The 90th percentile is the value associated with $u = 0.90$ on the graph, which is 750 cases.
c. There are 5 states that have a number of tuberculosis cases above the 90th percentile: Illinois, Florida, Texas, New York, and California.

3.75
a. There were 4 states that had number of AIDS, tuberculosis, and syphilis cases all above the 90th percentiles.
b. These four states were California, Florida, New York, and Texas. All four states are large states, but they are also associated with larger immigrant populations than other states.
c. The U.S. government should educate all people in all states about sexually-transmitted diseases. However, it appears that larger states and states with large immigrant populations need more education.

3.77
a. Mean = 9.905; Median = 9.544; $s = 1.550$
b. 25th percentile = 8.946; 50th percentile = median = 9.544; 75th percentile = 11.143
c. A boxplot and a histogram of the natural log of HIV-1 RNA levels are given here:

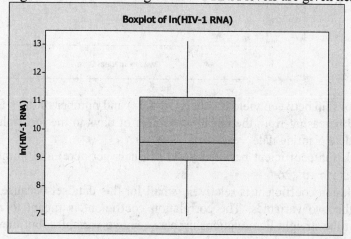

24 Chapter 3: Data Description

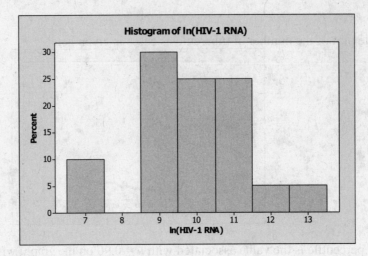

d. The natural logarithm transformation did result in a somewhat symmetric distribution, certainly more symmetric than the original distribution.

3.79
a. A scatterplot of yield (bushels per acre) versus number of days from the ideal planting date is given here:

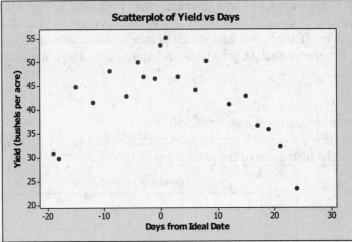

b. The relationship between yield (bushels per acre) and number of days from the ideal planting date is curved, increasing from the negative number of days to the ideal planting date and decreasing after the ideal planting date.
c. The correlation coefficient between yield (bushels per acre) and number of days from the ideal planting date is −0.226.
d. The correlation coefficient is relatively small for this data set because of the curved relationship between the two variables. The correlation coefficient is meant to measure the direction and strength of the straight line relationship between two variables, but these variables appear to have a quadratic relationship.

Chapter 4

Probability and Probability Distributions

4.1
  a. Subjective Probability
  b. Classical
  c. Relative Frequency
  d. Subjective Probability
  e. Subjective Probability
  f. Classical

4.3 The time period of a year is arbitrary. Probability is defined over a long (potentially infinite) time period. As we are only observing deaths over a fixed time period, there is bound to be differences simply due to the uncertainty inherent in probabilistic calculations. A decrease from one year to the next make be small enough to not be anything more than random fluctuation (What if we defined our 'year' to start or end in a different month? Would the results change?)

4.5 "Fair" means that each slot on the wheel has an equal chance of coming up and that the outcome of one spin does not impact the outcome on another.

4.7
  a. Let 00 to 79 represent positive outcomes and 80 to 99 represent negative outcomes.
  b. Let M be the number of sets of 30 two-digit numbers out of the 10,000 sets in which we have 24 or more "positive outcomes" (i.e., 24 or more two-digit numbers in the range 00 to 79). Then the probability would be approximated by M/10,000. This should be close to 0.5987.

4.9
  a. So long as we assume the numbers must be matched in sequence, the probability that your 4-digit number will be the winning number is $1/10000 = 0.0001$.
  b. Classical

4.11
  a. $A = \{\{1,1\}\}$; thus, $P(A) = 1/36 = 0.0278$
  b. $B = \{\{1,3\},\{2,2\},\{3,1\}\}$; thus, $P(B) = 3/36 = 0.0833$
  c. $C = \{\{1,1\},\{1,2\},\{1,3\},\{2,1\},\{2,2\},\{3,1\}\}$; thus, $P(C) = 6/36 = 0.1667$

4.13
  a. A and B are mutually exclusive.
  b. A and C are not mutually exclusive.
  c. B and C are not mutually exclusive.

4.15
a. All 3 pumps work.
b. At least one pump works (the plant is safe).
c. Not all three pumps are working (0, 1, or 2 pumps are working).
d. None of the pumps work (the plant isn't safe).

4.17
a. $P(\text{brown}) = 0.15$
b. $P(\text{red or green}) = 0.10 + 0.15 = 0.25$
c. $P(\text{not blue}) = 1 - 0.25 = 0.75$
d. $P(\text{both red and brown}) = 0$ (This is impossible!)

4.19
a. Because $P(A|B) \neq P(A)$, A and B are not independent.
b. Because $P(A|C) \neq P(A)$, A and C are not independent.
c. Because $P(B|C) \neq P(B)$, B and C are not independent.

4.21
a. The donor has type O blood given that the donor is white.
b. $P(T_1|W) = \dfrac{0.36}{0.802} = 0.4489$
c. $P(T_1|W) = 0.4489 \neq 0.462 = P(T_1)$, thus W and T_1 are not independent
d. $P(T_1 \cap W) = 0.36 \neq 0$, so W and T_1 are not mutually exclusive

4.23
a. $P(accepted) = 0.743$
b. $P(rejected) = 0.257$
c. $P(professional\ spouse) = 0.46$

4.25
a. $P(A|C) = 0.8486 \neq 0.743 = P(A)$; thus A and C are not independent
b. $P(A|D) = 0.9 \neq 0.743 = P(A)$; thus A and D are not independent
c. $1 - P(A|B) = 1 - \frac{276}{460} = 0.4 = \frac{184}{460} = P(\bar{A}|B)$ (They are equal.)
d. $1 - P(A|B) = 1 - \frac{276}{460} = 0.4 \neq 0.8648 = \frac{314+153}{370+170} = P(A|\bar{B})$ (They are not equal.)

4.27
a. P(both customers pay in full) = (0.70)(0.70) = 0.49
b. P(at least one of two customers pay in full) = 1 − P(neither customer pays in full)
   = $1 - (1 - 0.70)(1 - 0.70) = 1 - (0.30)^2 = 1 - 0.09 = 0.91$

4.29 Let D = event that the loan defaulted, R_1 = event that applicant is poor risk, R_2 = event that applicant is fair risk, R_3 = event that applicant is good risk.

$P(D) = 0.01,\quad P(R_1|D) = 0.30,\quad P(R_2|D) = 0.40,\quad P(R_3|D) = 0.30$

$P(\bar{D}) = 0.99,\quad P(R_1|\bar{D}) = 0.10,\quad P(R_2|\bar{D}) = 0.40,\quad P(R_3|\bar{D}) = 0.50$

$$P(D|R_1) = \frac{P(R_1|D)P(D)}{P(R_1|D)P(D) + P(R_1|\bar{D})P(\bar{D})} = \frac{(0.30)(0.01)}{(0.30)(0.01) + (0.10)(0.99)} = 0.0294$$

4.31

$$P(D_1|A_1) = \frac{P(A_1|D_1)P(D_1)}{P(A_1|D_1)P(D_1) + P(A_1|D_2)P(D_2) + P(A_1|D_3)P(D_3) + P(A_1|D_4)P(D_4)}$$

$$= \frac{(0.90)(0.028)}{(0.90)(0.028) + (0.06)(0.012) + (0.02)(0.032) + (0.02)(0.928)} = 0.5585$$

$$P(D_2|A_2) = \frac{(0.80)(0.012)}{(0.05)(0.028) + (0.80)(0.012) + (0.06)(0.032) + (0.01)(0.928)} = 0.4324$$

$$P(D_3|A_3) = \frac{(0.82)(0.032)}{(0.03)(0.028) + (0.05)(0.012) + (0.82)(0.032) + (0.02)(0.928)} = 0.5675$$

4.33 Let F = event fire occurs and T_i = event a type i furnace is in the home for $i = 1, 2, 3, 4$, where T_4 represents other types.

$$P(T_1|F) = \frac{P(F|T_1)P(T_1)}{P(F|T_1)P(T_1) + P(F|T_2)P(T_2) + P(F|T_3)P(T_3) + P(F|T_4)P(T_4)}$$

$$= \frac{(0.05)(0.30)}{(0.05)(0.30) + (0.03)(0.25) + (0.02)(0.15) + (0.04)(0.30)} = 0.40$$

4.35

$$P(A_2|B_1) = \frac{(0.17)(0.15)}{(0.08)(0.25) + (0.17)(0.15) + (0.10)(0.12)} = 0.4435$$

$$P(A_2|B_2) = \frac{(0.12)(0.15)}{(0.18)(0.25) + (0.12)(0.15) + (0.14)(0.12)} = 0.2256$$

$$P(A_2|B_3) = \frac{(0.07)(0.15)}{(0.06)(0.25) + (0.07)(0.15) + (0.08)(0.12)} = 0.2991$$

$$P(A_2|B_4) = \frac{(0.64)(0.15)}{(0.68)(0.25) + (0.64)(0.15) + (0.68)(0.12)} = 0.2762$$

4.37 Discrete. The number of helmet strangulations is something that we can count.

4.39
 a. Discrete
 b. Continuous

4.41
a. $P(y > 3) = 0.067 + 0.021 + 0.014 = 0.102$
b. $P(2 \leq y < 5) = 0.354 + 0.161 + 0.067 = 0.582$
c. $P(y > 4) = 0.021 + 0.014 = 0.035$

4.43 No, people may not answer the question.

4.45 Binomial experiment with y = number that over the legal limit, $n = 15$ and $\pi = 0.20$.
a. $P(y = 15) = 0.2^{15} = 3.28 \times 10^{-11} \approx 0$
b. $P(y = 6) = \binom{15}{6} 0.2^6 (1 - 0.2)^{15-6} = 0.043$
c. $P(y \geq 6) = \sum_{y=6}^{15} \binom{15}{y} 0.2^y (1 - 0.2)^{15-y} = 0.061$
d. $P(y = 0) = 0.8^{15} = 0.035$

4.47 No, the trials are not identical, because the patients and/or doctors may have different values for π.

4.49
a. $P(y = 4 | \mu = 2) = 0.0902$
b. $P(y = 4 | \mu = 3.5) = 0.188$
c. $P(y > 4 | \mu = 2) = 0.0527$
d. $P(1 \leq y < 4 | \mu = 2) = 0.7218$

4.51
a. Binomial experiment with $n = 2{,}500$ and $\pi = 0.001$. (From R)
$$P(y > 5) = 1 - P(y \leq 5) = 1 - \sum_{x=0}^{5} \binom{2500}{x} (0.001)^x (0.999)^{2500-x} = 0.0419$$
b. We need to assume that Internet hits are independent and that the probability of a sale remains constant.
c. Let $\mu = np = 2500 * 0.001 = 2.5$
$$P(y > 5) = 1 - P(y \leq 5) = 0.042$$
d. The Poisson approximation is very close to the actual binomial calculation.

4.53
a. $P(0 < z < 1.3) = 0.9032 - 0.5000 = 0.4032$
b. $P(0 < z < 2.7) = 0.9965 - 0.5000 = 0.4965$

4.55
a. $P(-2.5 < z < -1.2) = 0.1151 - 0.0062 = 0.1089$
b. $P(-1.3 < z < -0.7) = 0.2420 - 0.0968 = 0.1452$

4.57 $P(z < 1.23) = 0.8907$

4.59 $z_0 = 0$

4.61 $z_0 = 2.36$

Chapter 4: Probability and Probability Distributions

4.63 $z_0 = 1.645$

4.65 y is Normally distributed with μ=250 and σ=50
  a. $P(y > 250) = P\left(z > \frac{250-250}{50}\right) = P(z > 0) = 0.5$
  b. $P(y > 150) = P\left(z > \frac{150-250}{50}\right) = P(z > -2) = 0.9772$
  c. $P(150 < y < 350) = P\left(\frac{150-250}{50} < z < \frac{350-250}{50}\right) = P(-2 < z < 2)$
  $= 0.9772 - 0.0228 = 0.9544$
  d. $P(250 - k < y < 250 + k) = 0.6$
  From Table 1 we find, $P(-0.842 < z < 0.842) = 0.6$
  Thus, $\frac{250+k-250}{50} = 0.842 \Rightarrow k = 42.1$

4.67
  a. $z = 2.326$
  b. $z = -1.96$
  c. $z = 1.96$

4.69
  a. $P(y > 50) = P\left(z > \frac{50-39}{6}\right) = P(z > 1.83) = 1 - 0.9664 = 0.0336$
  b. $P(y > 55) = P\left(z > \frac{55-39}{6}\right) = P(z > 2.67) = 1 - 0.9962 = 0.0038$
  Since the probability is so low, we would conclude that the voucher has been lost.

4.71
  a. $P(y < 200) = P\left(z < \frac{200-150}{35}\right) = P(z < 1.43) = 0.9236$
  b. $P(y > 100) = P\left(z > \frac{100-150}{35}\right) = P(z > -1.43) = 0.9236$
  c. $P(100 < y < 200) = P\left(\frac{100-150}{35} < z < \frac{200-150}{35}\right) = P(-1.43 < z < 1.43) = 0.8472$

4.73 If the interviewer visits homes only from 8 AM to 5 PM, then most homes with occupants employed would be excluded, hence a biased sample.

4.75 Start at column 2 line 1. We obtain 150, 729, 611, 584, 255, 465, 143, 127, 323, 225, 483, 368, 213, 270, 062, 399, 695, 540, 330, 110, 069, 409, 539, 015, 564. These would be the women selected for the study.

4.77 The sampling distribution would have a mean of 60 and a standard deviation of $5/\sqrt{16} = 1.25$. If the population distribution is somewhat mound-shaped, then the sampling distribution of $\bar{y}$ should be approximately mound-shaped. In this situation, we would expect approximately 95% of the possible values of $\bar{y}$ to lie in $60 \pm 2(1.25) = (57.5, 62.5)$.

4.79

a. $P(900 < \bar{x} < 960) = P\left(\frac{900-930}{130/\sqrt{20}} < z < \frac{960-930}{130/\sqrt{20}}\right) = P(-1.03 < z < 1.03)$
$= 0.8485 - 0.1515 = 0.697$

b. $P(\bar{x} > 960) = P\left(z > \frac{960-930}{130/\sqrt{20}}\right) = P(z > 1.03) = 1 - 0.8485 = 0.1515$

c. $P(z < 1.28) = 0.9 \Rightarrow 1.28 = \frac{\bar{x}-930}{130/\sqrt{20}} \Rightarrow \bar{x} = 967.208$

4.81

a. $P(z > 1.28) = 0.10 \Rightarrow k = 125 + 1.28(32) = 165.96 \Rightarrow$ facility size should be at least 166.

b. $P(z > 0.524) = 0.30 \Rightarrow k = 125 + 0.524(32) = 141.8 \Rightarrow$ facility size should be at least 142.

4.83

a. $P(y > 2.7) = P\left(z > \frac{2.7-2.1}{0.3}\right) = P(z > 2) = 0.0228$

b. $P(z > 0.6745) = 0.25 \Rightarrow y_{0.75} = 2.1 + 0.6745(0.3) = 2.3$

c. Let μ_N be the new value of the mean. We need $P(y > 2.7) = 0.05$. This implies that the 95th percentile must by 2.7. That is,
$2.7 = \mu_N + 1.645(0.3) \Rightarrow$
$\mu_N = 2.7 - 0.4935 = 2.2$
Therefore, the standard has already been met without any changes.

4.85 Individual baggage weight has $\mu = 95$ and $\sigma = 35$. Total weight has mean $n\mu = 200(95) = 19,000$ and standard deviation $\sqrt{n}\sigma = \sqrt{200}(35) = 494.97$.

Therefore, $P(y > 20,000) = P\left(z > \frac{20,000-19,000}{494.97}\right) = P(z > 2.02) = 0.0217$.

4.87

a. $\mu = n\pi = (10,000)(0.001) = 10$

b. $\sigma = \sqrt{(10,000)(0.001)(0.999)} = 3.16$

$P(y < 4) = P(y \leq 3) \approx P\left(z \leq \frac{3.5-10}{3.16}\right) = P(z \leq -2.06) = 0.0197$

c. $P(y > 2) = 1 - P(y \leq 2) \approx 1 - P\left(z \leq \frac{2.5-10}{3.16}\right) = 0.9911$

4.89

a. $P(4 \leq y \leq 6) = \sum_{i=4}^{6} \binom{10}{i}(0.5)^i(0.5)^{10-i} = 0.65625$

b. $\mu = n\pi = (10)(0.5) = 5$; $\sigma = \sqrt{(10)(0.5)(0.5)} = 1.58$;

$P(4 \le y \le 6) = P(y \le 6) - P(y \le 3) \approx P\left(z \le \frac{6-5}{1.58}\right) - P\left(z \le \frac{3-5}{1.58}\right)$

$= P(z \le 0.63) - P(z \le -1.27) = 0.6337$

This is fairly close to the actual probability.

4.91
a. Using the normal approximation with $\mu = n\pi = 10,000(0.15) = 1500$;
$\sigma = \sqrt{(10,000)(0.15)(0.85)} = 35.7071$

$P(y < 1430) \approx P\left(z < \frac{1429.5 - 1500}{35.7071}\right) = P(z < -1.97) = 0.0244$

b. $P(y > 1600) \approx P\left(z > \frac{1600.5 - 1500}{35.7071}\right) = P(z > 2.81) = 1 - 0.9975 = 0.0025$

4.93
a. A boxplot, relative frequency histogram, and normal probability plot appear here.

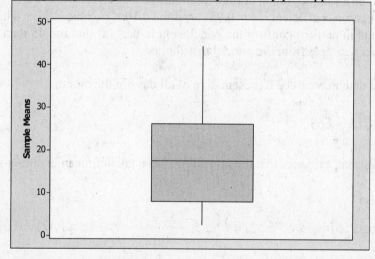

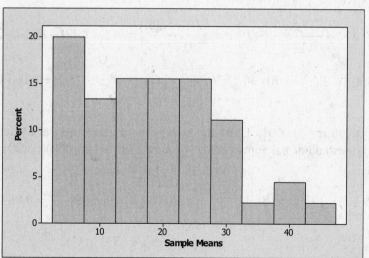

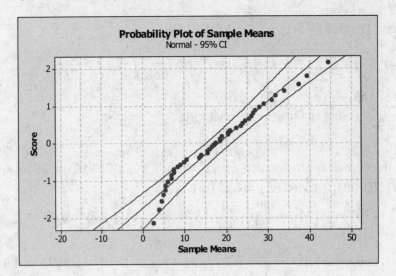

The boxplot and the histogram appear slightly right skewed. The normal probability plot has a slight curve to it. However, all but a few points are floes to the fitted line. Therefore, the 45 points appear to be a sample from a normal distribution.

b. The correlation coefficient is $r = 0.981$, and the $0.10 < p\text{-value} < 0.25$. This indicates a good fit.
c. The results in part (b) confirm the conclusion in part (a) that the 45 data values do not appear to be a random sample from the normal distribution.

4.95 No. The last date may not be representative of all days in the month.

4.97 $\sigma_{\bar{y}} = 10.2/\sqrt{15} = 2.63$

4.99 No, there is strong evidence that the new fabric has a greater mean breaking strength.

4.101

a. $P(y > \log(250)) = P(y > 5.52) = P\left(z > \dfrac{5.52 - 5.35}{0.12}\right) = P(z > 1.42) = 0.0765 \Rightarrow 7.65\%$

b. $P(\log(150) < y < \log(250)) = P(5.01 < y < 5.52)$
$= P\left(z < \dfrac{5.52 - 5.35}{0.12}\right) - P\left(z < \dfrac{5.01 - 5.35}{0.12}\right) = P(z < 1.42) - P(z < -2.83) = 0.9200 \Rightarrow 92\%$

$P(y > \log(300)) = P(y > 5.70) = P\left(z > \dfrac{5.70 - 5.35}{0.12}\right) = P(z > 2.92) = 0.0018 \Rightarrow 0.18\%$

4.103 $n = 20,000$ and $\pi = 0.0001$. There are two possible outcomes, and each birth is an independent event. We cannot use the normal approximation because $n\pi = (20,000)(0.0001) = 2 < 5$. We can use the binomial formula:

$$P(y \geq 1) = 1 - P(y = 0) = 1 - \binom{20000}{0}(0.0001)^0 (0.9999)^{20000} = 0.8647$$

4.105
a. The normal probability plot is given here.

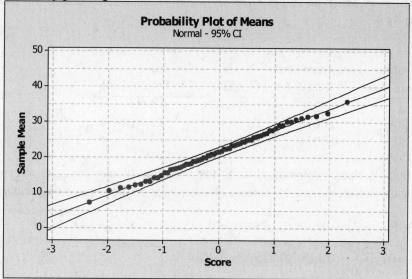

Note that the plotted points deviate only slightly from the straight line. Thus, even though the sample size is very small the normal approximation fits quite well. The reason for the closeness of the approximation is that the population has a somewhat symmetric distribution.

b. Both the population mean and the mean of the 70 $\bar{y}$'s are equal to 21.5.

4.107
a. The normal probability plot is given here.

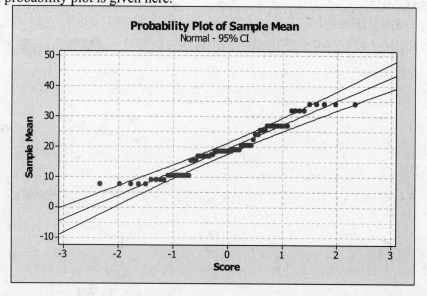

Note that the plotted points deviate considerably from the straight line. Thus, the sampling distribution is not approximated very well by a normal distribution. If the sample size was much larger than 4, the approximation would be greatly improved.

b. The population median equals $\frac{12+25}{2} = 18.5$ whereas the mean of the 70 values of the sample median is 19.536. The values differ by a significant amount due to the fact that the sample size was only 4.

4.109

a.-c. The probabilities are given here.

Sample Size	$P(\bar{y} \geq 105)$	$P(\bar{y} \leq 95)$	$P(95 \leq \bar{y} \leq 105)$
5	0.2280	0.2280	0.5439
20	0.0680	0.0680	0.8640
80	0.0014	0.0014	0.9971

The sample mean's chance of being close to the population mean increases very rapidly as the sample size increases. Of course, the rate of increase also depends on the population standard deviation, σ.

4.111 Let y be binomial with $n = 12$, $\pi = 1/8 = 0.125$.

a. $P(y=0) = \binom{12}{0}(0.125)^0(0.875)^{12} = 0.2014$

b. $P(y \geq 2) = 1 - P(y=0) - P(y=1) = 0.4533$

c. $P(y \leq 4) = \sum_{i=0}^{4} \binom{12}{i}(0.125)^i(0.875)^{12-i} = 0.9887$

4.113

a. The Poisson probability model might be appropriate.
b. $\mu = 6(2.4) = 14.4$
c. $P(y=0) = 0.0000$ (from Table 15)
d. $P(y \geq 2) = 1 - P(y=0) - P(y=1) = 1 - 0.0000 - 0.0000 \approx 1.0000$

Chapter 5

Inferences about Population Central Values

5.1
  a. The population of interest consists of adult residents of the county who are eligible for jury duty.
  b. We could take a simple random sample from a list of all registered voters in the county.

5.3
  a. The population is the face masks produced by the manufacturer during a selected period of time.
  b. Testing a hypothesis.

5.5
  a. All coffee produced by this production line (or company if the line is representative of the company).
  b. $110 \pm (1.96)\frac{7.1}{\sqrt{50}} = 110 \pm 1.97 = (108.0, 112.0)$
  c. We are 95% confident that the average caffeine content is between 108 and 112 milligrams.

5.7
  a. $(0.05)(720) = 36$
  b. $(0.05)(720) = 36$
  c. $(0.01)(720) = 7.2$, or about 7

5.9
  a. $2.8 \pm (1.96)\frac{1.3}{\sqrt{200}} = 2.8 \pm 0.18 = (2.62 \, yrs, 2.98 \, yrs)$
  b. As the records were only from the courthouse of a single city, the confidence interval only really applies to two-time offenders from that city.
  c. It would be unadvisable to generalize the results to a larger population other than two-time offenders from that city.

5.11
  a. $n = \frac{(2.58)^2(0.7)^2}{(0.15)^2} = 144.962 \Rightarrow 145$
  b. $n = \frac{(1.96)^2(0.7)^2}{(0.25)^2} = 30.12 \Rightarrow 31$
  c. $n = \frac{(2.81)^2(0.7)^2}{(0.25)^2} = 61.91 \Rightarrow 62$
  d. For a fixed desired width, an increase in confidence level will lead to an increase in sample size (and vice versa).
  e. For a fixed confidence level, an increase in desired width will lead to a decrease in sample size (and vice versa).

5.13 $\hat{\sigma} = 13, E = 3, \alpha = 0.01 \Rightarrow n = \frac{(2.58)^2(13)^2}{(3)^2} = 124.99 \Rightarrow n = 125$

5.15

a. $\hat{\sigma} = 625$; $E = 100$; $\alpha = 0.01 \Rightarrow n = \dfrac{(2.58)^2(625)^2}{(100)^2} = 260.016 \Rightarrow n = 261$

b. The 95% level of confidence implies that there will be a 1 in 20 chance, over a large number of samples, that the confidence interval will not contain the population average rent. The 99% level of confidence implies that there is only a 1 in 100 chance of not containing the average rent. Thus, we would increase the odds of not containing the true average five-fold.

5.17

a. As the value of μ decreases from $\mu_0 = 26$, the power increases steadily to 100%.

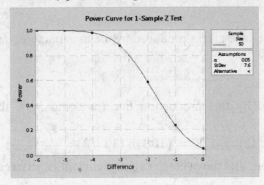

b. The power increase to 1 slower for smaller α.

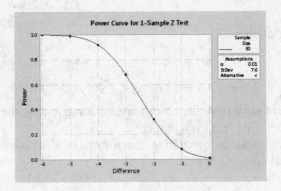

c. The power increases as μ decreases more slowly when n is reduced.

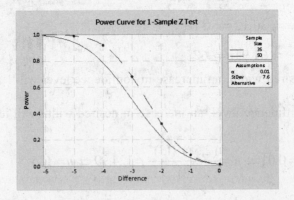

5.19
  a. Answers will vary.
  b. $PWR(45) = P\left(z > 1.96 - \frac{|43-45|}{4/\sqrt{25}}\right) = P(z > -0.54) = 0.7054$
  c. On average, 70.54 tests will result in a correct decision.

5.21 $H_0: \mu \leq 2$ vs $H_a: \mu > 2$, $\bar{y} = 2.17$, $s = 1.05$, $n = 90$

  a. $z = \dfrac{2.17 - 2}{1.05/\sqrt{90}} = 1.54 < 1.645 = z_{0.05} \Rightarrow$ fail to reject H_0. The data does not support the hypothesis that the mean has been decreased from 2.

  b. $\beta(2.1) = P\left(z \leq 1.645 - \dfrac{|2 - 2.1|}{1.05/\sqrt{90}}\right) = P(z \leq 0.74) = 0.7704$

5.23 $n = \dfrac{(4)^2 (1.645 + 1.28)^2}{(15 - 15.3)^2} = 1521$

5.25 $z = \dfrac{542 - 525}{76/\sqrt{100}} = 2.24 > 1.645 = z_{0.05} \Rightarrow$ Reject H_0

There is sufficient evidence to conclude that the mean has been increased above 525.

5.27 $H_0: \mu \leq 30$ vs $H_a: \mu > 30$
$\alpha = 0.05$, $n = 37$, $\bar{y} = 37.24$, $s = 37.12$

  a. $z = \dfrac{37.24 - 30}{37.12/\sqrt{37}} = 1.19 < 1.645 = z_{0.05} \Rightarrow$ Fail to reject H_0

  There is not sufficient evidence to conclude that the mean lead concentration exceeds 30 mg/kg dry weight.

  b. $\beta(50) = P\left(z \leq 1.645 - \dfrac{|30 - 50|}{37.12/\sqrt{37}}\right) = P(z \leq -1.63) = 0.0513$

  c. No, the data values are not very close to the straight-line in the normal probability plot. Most of the deviation is likely due to the outlier at 210.
  d. No, since there is a substantial deviation from a normal distribution, the sample size should be somewhat larger to use the z-test.

5.29 $p - value = P(z \leq -1.08) = 0.1401 > 0.10 = \alpha$

No, there is still not significant evidence that the mean is less than 35 at the 0.10 level.

5.31 In testing $H_0: \mu \geq 21.7$ versus $H_a: \mu < 21.7$,

  p-value $= P\left(z \leq \dfrac{18.8 - 21.7}{15.3/\sqrt{90}}\right) = P(z \leq -1.80) = 0.0359 < 0.05 = \alpha$

We would now conclude there is significant evidence that the mean is less than 21.7.

5.33 $H_0: \mu = 1.6$ vs $H_a: \mu \neq 1.6$

$n = 36, \bar{y} = 2.2, s = 0.57, \alpha = 0.05$

p-value = $2P\left(z \geq \frac{|2.2-1.6|}{0.57/\sqrt{36}}\right) = 2P(z \geq 6.32) < 0.0001 < 0.05 = \alpha$

Yes, there is significant evidence that the mean time delay differs from 1.6 seconds.

5.35

a. $p - value = P\left(t \geq \frac{10.1-9}{\frac{3.1}{\sqrt{17}}}\right) = P(t \geq 1.46) \Rightarrow 0.05 < p - value < 0.10$

Since p-value > 0.05, there is not significant evidence that the mean is greater than 9.

b. The level of significance is 0.0818 (from computer).

5.37 $H_0: \mu \leq 80$ vs $H_a: \mu > 80$

$n = 20, \bar{y} = 82.05, s = 10.88$; Reject H_0 if $t \geq 1.729$

$t = \frac{82.05 - 80}{10.88/\sqrt{20}} = 0.84 < 1.729$

Thus, fail to reject H_0 and conclude the data does not support the hypothesis that the mean reading comprehension is greater than 80.

The level of significance is given by p-value = $P(t \geq 0.84) \approx 0.20$.

5.39 $n = 15, \bar{y} = 31.47, s = 5.04$

a. $31.47 \pm 2.977 \frac{5.04}{\sqrt{15}} = 31.47 \pm 3.87 = (27.6, 35.34) \Rightarrow$ We are 99% confident that the mean miles driven is between 27,600 and 35,340.

b. $H_0: \mu \geq 35$ vs $H_a: \mu < 35$; Reject H_0 if $t \leq -2.624$

$t = \frac{31.47 - 35}{5.04/\sqrt{15}} = -2.71 < -2.624$

Thus, reject H_0 and conclude the data supports the hypothesis that the mean miles driven is less than 35,000 miles.

The level of significance is given by
p-value = $P(t \leq -2.71) \Rightarrow 0.005 <$ p-value < 0.01.

5.41

a. $4.95 \pm 2.365 \frac{0.45}{\sqrt{8}} = 4.95 \pm 0.38 = (4.57, 5.33)$ We are 95% confident that the mean dissolved oxygen level for the population is between 4.57 and 5.33 ppm.

b. There is inconclusive evidence that the mean is less than 5 ppm since the CI contains values both less and greater than 5.

c. $H_0: \mu \geq 5$ vs $H_a: \mu < 5$

p-value = $P\left(t \leq \dfrac{4.95-5.0}{0.45/\sqrt{8}}\right) = P(t \leq -0.31) \Rightarrow 0.25 < $ p-value < 0.40

(Using a computer program, p-value = 0.3828.) Fail to reject H_0 and conclude the data does not support that the mean is less than 5 ppm.

5.43 Answers will vary.

5.45 Answers will vary.

5.47
   a. $C_{0.05,12} = 2 \rightarrow L_{0.025} = 3$ and $U_{0.025} = 10 \rightarrow (y_{(3)}, y_{(10)})$
   b. $C_{0.05,12} \approx 6 - 1.96\sqrt{3} = 2.61 \approx 2 \rightarrow L_{0.025} = 3$ and $U_{0.025} = 10 \rightarrow (y_{(3)}, y_{(10)})$
   c. Even with a small sample size ($n = 12$), the normal approximation agrees with the actual interval. Without knowing the raw data, the interval is based on the order statistics.

5.49 Reject if $B \geq 25 - 6 = 19$

5.51
   a. The normal probability plot and boxplot are given here:

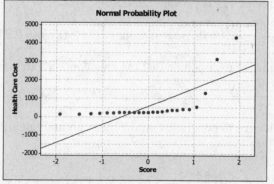

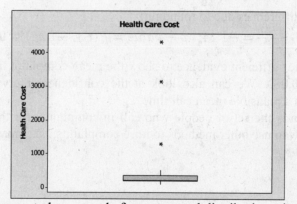

The data set does not appear to be a sample from a normal distribution, since a large proportion of the values are outliers as depicted in the boxplot and several points are a considerable distance from the line in the normal probability plot. The data appears to be from an extremely right-skewed distribution.

b. Because of the skewness, the median would be a better choice than the mean.
c. (208, 342) We are 95% confident that the median amount spent on healthcare by the population of hourly workers is between $208 and $342.
d. Reject $H_0 : M \leq 400$ if $B \geq 25 - 7 = 18$.

We obtain B = 4. Since 4 < 18, do not reject $H_0 : M \leq 400$. The data fails to demonstrate that the median amount spent on health care is greater than $400.

5.53
a. Reject $H_0 : M \leq 0.25$ in favor of $H_a : M > 0.25$ at level $\alpha = 0.01$ if $B \geq 25 - 6 = 19$. From the differences, $y_i - 0.25$, we obtain B = 18 positive values. Thus, we fail to reject H_0 and conclude that the data does not support a median increase in reaction time of more than 0.25 seconds.
b. Weight of driver, experience and age of driver, amount of sleep in previous 24 hours, etc.

5.55
a. Using $\alpha = 0.05$, reject $H_0 : M \leq 10$ in favor of $H_a : M > 10$ if $B \geq 9$.
Fund A: B = 6 $\Rightarrow$ fail to reject H_0. There is not sufficient evidence to conclude that the median annual rate of return is greater than 10%.
Fund B: B = 6 $\Rightarrow$ fail to reject H_0. There is not sufficient evidence to conclude that the median annual rate of return is greater than 10%.
b. Using $\alpha = 0.05$, reject H_0 if p-value ≤ 0.05.

Fund A: $t = \dfrac{13.65 - 10}{15.87 / \sqrt{10}} = 0.73 \Rightarrow$ p-value $= P(t \geq 0.73) = 0.242 \Rightarrow$ fail to reject H_0. There is not sufficient evidence to conclude that the mean annual rate of return is greater than 10%.

Fund B: $t = \dfrac{16.56 - 10}{16.23 / \sqrt{10}} = 1.28 \Rightarrow$ p-value $= P(t \geq 1.28) = 0.1163 \Rightarrow$ fail to reject H_0. There is not sufficient evidence to conclude that the mean annual rate of return is greater than 10%.

5.57
a. $\bar{x} \pm t \dfrac{s}{\sqrt{n}} \Rightarrow 26.218 \pm 2.01 \dfrac{4.461}{\sqrt{50}} \Rightarrow (24.95, 27.49)$ (Note: this was solved on a computer so there will be slight differences due to rounding)
b. $t = \dfrac{\bar{x} - \mu}{\frac{s}{\sqrt{n}}} = \dfrac{26.218 - 27}{\frac{4.461}{\sqrt{50}}} = -1.24; \quad p-value = P(t \leq -1.24) = 0.1104$

We do not have sufficient evidence to show the mean complaint time was less than 27 minutes as $p-value > 0.05$. We can also look at the confidence interval and notice that 27 is in the interval so it is a plausible mean call time.
c. The population is the set of people who call in complaints for this Internet company (it doesn't include those who use other methods to raise complaints, i.e., email, chat, etc.).

5.59
a. A normal probability plot and boxplot are given here:

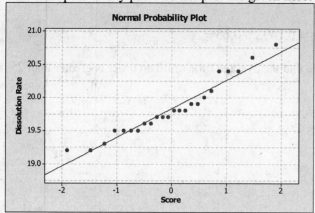

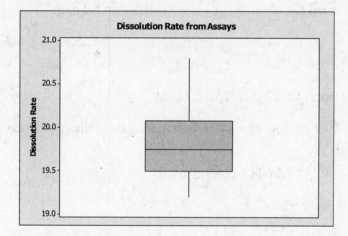

The plots indicate that the data was selected from a population having a distribution that is somewhat skewed to the right, but only slightly, since there are no outliers indicated on the boxplot.

b. $\bar{y} = 19.83$; 99% CI: $19.83 \pm 2.807 \dfrac{0.43}{\sqrt{24}} = 19.83 \pm 0.25 = (19.58, 20.08)$ We are 99% confident that the average dissolution rate is between 19.58 and 20.08 minutes.

c. Test $H_0: \mu \geq 20$ versus $H_a: \mu < 20$.

$n = 24, \bar{y} = 19.83, s = 0.43, \alpha = 0.01$

p-value = $P\left(t \leq \dfrac{19.83 - 20}{0.43/\sqrt{24}}\right) = P(t \leq -1.94) = 0.0324 > 0.01 = \alpha$

No, there is not sufficient evidence to conclude that the average dissolution rate is less than 20 minutes.

d. From Table 3 with $d = \dfrac{|19.6 - 20|}{0.43} = 0.93$, $df = 23$, $\alpha = 0.01$, we obtain $\beta(19.6) = 0.025$ (actually, this value was obtained from a computer program since this degree of accuracy could not be obtained from Table 3).

42 Chapter 5: Inferences about Population Central Values

5.61 $H_0: \mu \leq 25$ versus $H_a: \mu > 25$
$n = 15, \bar{y} = 28.20, s = 11.44, \alpha = 0.05$
p-value = $P\left(t \geq \dfrac{28.20-25}{11.44/\sqrt{15}}\right) = P(t \geq 1.08) = 0.1492 > 0.05 = \alpha$
No, there is not sufficient evidence to conclude that the average time to fill an order is greater than 25 minutes.

5.63
  a. $H_0: \mu \geq 5.2$ versus $H_a: \mu < 5.2$
  $n = 50, \bar{y} = 5.0, s = 0.70, \alpha = 0.05$
  $z = \dfrac{5.0-5.2}{0.7/\sqrt{50}} = -2.02$
  Reject H_0 if $z \leq -1.645$
  b. Reject H_0 and conclude that the mean dissolved oxygen count is less than 5.2 ppm.

5.65
  a. $\bar{y} = 30.514, s = 12.358, n = 35$
  95% CI: $30.514 \pm 2.032 \dfrac{12.358}{\sqrt{35}} = 30.514 \pm 4.245 = (26.27, 34.76)$
  We are 95% confident that the interval 26.27 to 34.76 minutes captures the population mean exercise capacity.
  b. 99% CI: $30.514 \pm 2.728 \dfrac{12.358}{\sqrt{35}} = 30.514 \pm 5.698 = (24.81, 36.21)$
  The 99% CI is somewhat wider than the 95% CI.

5.67 $n = 50, \bar{y} = 75, s = 15$
95% CI on μ: $75 \pm 2.010 \dfrac{15}{\sqrt{50}} = 75 \pm 4.3 = (70.7, 79.3)$

5.69 $\hat{\sigma} = \dfrac{400-40}{4} = 90 \Rightarrow n = \dfrac{(90)^2(1.96)^2}{(10)^2} = 311.2 \Rightarrow n = 312$

5.71
  a. 95% CI on median: $(y_{(6)}, y_{(15)}) = (3.59, 7.76)$
  We are 95% confident that the median percentage of SiO_2 is between 3.59% and 7.76%.
  b. Using $\alpha = 0.05$, reject $H_0: M = 6.2$ in favor of $H_a: M \neq 6.2$ if $B \leq 5$ or $B \geq 15$.
  We obtain B = 7. Since 5 < 7 < 15, do not reject $H_0: M \leq 6.2$. The data fails to demonstrate that the median percentage of SiO_2 is different from 6.2%.

5.73 From the pilot study, $\hat{\sigma} = 9.8$, therefore, $n = \dfrac{(9.8)^2(2.33+1.28)^2}{(4)^2} = 78.23 \Rightarrow n = 79$

5.75
- a. $H_0: \mu_{after} - \mu_{before} \leq 0$ versus $H_a: \mu_{after} - \mu_{before} > 0$ which implies

  $H_0: \mu_{after} \leq \mu_{before}$ versus $H_0: \mu_{after} > \mu_{before}$

  $t = \dfrac{2.11 - 0}{7.54/\sqrt{10}} = 0.88 \Rightarrow$ p-value $= P(t \geq 0.88) = 0.2009 > 0.05$

  There is not significant evidence that the mean change in mpg is greater than 0, i.e., that the mean mpg has been increased after installing the device.

- b. $2.11 \pm 1.833 \dfrac{7.54}{\sqrt{10}} = 2.11 \pm 4.37 = (-2.26, 6.48)$

  Using the decision rule: Reject $H_0: \mu \leq \mu_0$ in favor of $H_a: \mu > \mu_0$ if μ_0 is less than the lower limit of the CI, we have that since 0 is greater than the lower limit of the CI (-2.26), we fail to reject H_0 and conclude that there is not significant evidence that the difference in the average mpg is greater than 0.

Chapter 6

Inferences Comparing Two Population Central Values

6.1
- a. The population of interest is *Distichlis spicata*, a flora of particular interest in both contaminated and uncontaminated areas of the marsh.
- b. New growth on existing plants, pH of the soil
- c. Number the tracts from 1 to N, where N is the total number of tracts. Use a computer to generate n random numbers and select the tracts associated with those random numbers.
- d. Is the proportion of flora with new growth the same in the spill sites and control sites? Is the pH of the soil different between spill sites and control sites?

6.3 $H_0: \mu_1 \geq \mu_2 - 2.3$ versus $H_A: \mu_1 < \mu_2 - 2.3$
Reject if $t < -2.449$ (df = 32)

$$s_p = \sqrt{\frac{(13-1)7.23^2 + (21-1)6.98^2}{13+21-2}} = \sqrt{50.05} = 7.07$$

$$t = \frac{(50.3-58.6)-2.3}{7.07\sqrt{\frac{1}{13}+\frac{1}{21}}} = -4.24 < -2.449 \Rightarrow \text{reject } H_0$$

The data provide significant evidence that μ_1 is less than $\mu_1 - 2.3$.

6.5
- a. $H_0: \mu_{26} \geq \mu_5$ versus $H_a: \mu_{26} < \mu_5$

  $\bar{y}_{26} = 165.8, s_{26} = 14.77, n_{26} = 6$, $\bar{y}_2 = 378.5, s_2 = 23.96, n_2 = 6$, df = 10

  Reject H_0 if $t \leq -1.812$

  $$s_p = \sqrt{\frac{(5)14.77^2 + (5)23.96^2}{10}} = \sqrt{396.11725} = 19.9$$

  $$t = \frac{165.8 - 378.5 - 0}{19.9\sqrt{1/6 + 1/6}} = -18.513 < -1.812 \Rightarrow \text{reject } H_0$$

  There is significant evidence that μ_{26} is less than μ_5, with p-value < 0.0005.

- b. The sample sizes are too small to evaluate the normality condition but the sample variances are fairly close considering the sample sizes. We would need to check with the experimenter to determine if the two random samples were independent.

- c. $165.8 - 378.5 \pm 2.228(19.9)\sqrt{\frac{1}{6}+\frac{1}{6}} = -212.7 \pm 25.6 = (-238.3, -187.1)$

  A 95% CI on the mean difference is $(-238.3, -187.1)$, which indicates that the average warm temperature rat blood pressure is between 187 and 239 units lower than the average 5°C rat blood pressure.

6.7
- a. $H_0: \mu_U \geq \mu_S$ versus $H_a: \mu_U < \mu_S$; p-value < 0.0005 ⇒ The data provide sufficient evidence to conclude that successful companies have a lower percentage of returns than unsuccessful companies.
- b. The boxplots indicate that both data sets appear to be from normally distributed distributions; however, the successful data set indicates a higher variability than the unsuccessful.

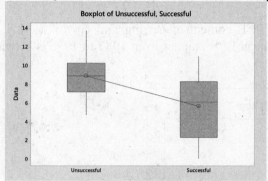

- c. The results from Minitab are below

```
Two-sample T for Unsuccessful vs Successful

                N    Mean   StDev   SE Mean
Unsuccessful    50   8.97   2.20    0.31
Successful      50   5.72   3.24    0.46


Difference = μ (Unsuccessful) - μ (Successful)
Estimate for difference:  3.251
95% lower bound for difference:  2.329
T-Test of difference = 5 (vs >): T-Value = -3.16  P-Value = 0.999  DF = 86
```
There is not significant evidence that the percentage for successful businesses returned goods is 5% less than that of unsuccessful businesses.
- d. `95% CI for difference:  (2.149, 4.353)`

6.9.
- a. All sets of 10,000 scm of air from the South Pole.
- b. The test run below suggest significant evidence of a difference in average g/scm of the two metals.

Two-Sample T-Test and CI

```
Sample   N   Mean   StDev   SE Mean
1        35  1.00   2.21    0.37
2        35  17.0   12.7    2.1


Difference = μ (1) - μ (2)
Estimate for difference:  -16.00
95% CI for difference:  (-20.40, -11.60)
T-Test of difference = 5 (vs ≠): T-Value = -9.67  P-Value = 0.000  DF = 36
```

- c. The level of significance (p-value) is approximately zero.
- d. As seen above, the 95% confidence interval for the difference is $(-20.40, -11.60)$

6.11

a. $H_0: \mu_{96} \geq \mu_{82}$ versus $H_a: \mu_{96} < \mu_{82}$; $t = \dfrac{15.52 - 54.33}{\sqrt{\dfrac{(5.96)^2}{13} + \dfrac{(15.65)^2}{13}}} = -8.35$; df = 15;

p-value $< 0.0005 \Rightarrow$ reject H_0. The data provide sufficient evidence that there has been a decrease in mean PCB content.

b. A 95% CI on the difference in the mean PCB content of herring gull eggs is (−48.4, −29.2), which would indicate that the decrease in mean PCB content from 1982 to 1996 is between 29.2 and 48.4.

c. The boxplots are given here:

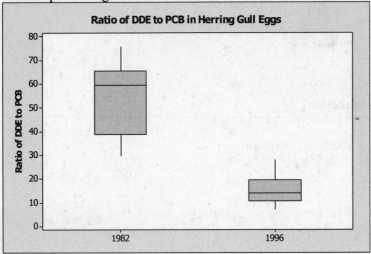

The boxplots of the PCB data from the two years both appear to support random samples from normal distributions, although the 1982 data is somewhat skewed to the left. The variances for the two years are substantially different; hence the separate-variance t-test was applied in part (a).

d. Since the data 1982 and 1996 were collected at the same sites, there may be correlation between the two years. There may also be spatial correlation depending on distance between sites.

6.13

a. All lower-level managers at this firm.

b. The p-value of the test below is very large so there is not significant evidence that males make more the five units higher in average bonus pay.

```
Two-sample T for F vs M

    N   Mean   StDev   SE Mean
F   24  8.53   1.19    0.24
M   36  9.68   1.00    0.17

Difference = μ (F) - μ (M)
Estimate for difference:  -1.150
95% upper bound for difference:  -0.654
T-Test of difference = -5 (vs <): T-Value = 13.06  P-Value = 1.000  DF = 43
```

c. The p-value is approximately 1.

d. Using unequal variances, the 95% CI is: (-1.744, -0.556)

6.15
   a. Reject H_0 if T > 87 or T < 49; T = 91.5 > 87 $\Rightarrow$ reject H_0. There is sufficient evidence to conclude that median for Treatment 1 is different from the median of Treatment 2.
   b. We have 64 differences. For $\alpha = 0.025, T_U = 87$
   $C_{0025} = \frac{8(16+8+1)}{2} + 1 - 87 = 14$ so
   $\Delta_L = D_{(C_{0025})} = D_{14} = 0.2; \Delta_U = D_{64+1-C_{0025}} = D_{61} = 15$
   95% CI: (0.2, 1.5)

6.17
   a. The cable repairmen who complete service calls for this company.
   b. The output below shows we do not have significant evidence at the 0.05 level to conclude the new method is more efficient (barely).

   Two-Sample T-Test and CI: Treatment, Control

   ```
   Two-sample T for Treatment vs Control

              N    Mean   StDev   SE Mean
   Treatment  18   88.89   9.33     2.2
   Control    18   102.7   33.0     7.8


   Difference = μ (Treatment) - μ (Control)
   Estimate for difference:  -13.79
   95% upper bound for difference:  0.19
   T-Test of difference = 0 (vs <): T-Value = -1.71   P-Value = 0.052   DF = 19
   ```
   c. The p-value is 0.052.
   d. A two-sided 95% CI for the mean difference is 95% CI for difference: (-30.71, 3.14)
   e. One can use a confidence interval approach, t test, or Wilcoxon test. Each of the three tests has its merits and disadvantages. The treatment group appears to be fairly symmetric, but with high and low outliers. The control group is skewed to the right with high and low outliers. With the small sample size and possible lack of normality, we should use the Wilcoxon rank sum test. However, we should decide to use one test and not run all three tests. Running all three tests might tempt us to use the one that has the results that we want (if the results differ).

6.19
   a. The heavy-tailness (Cauchy) has the greatest impact on the significance level (considerably lower than $\alpha = 0.05$).
   b. The Wilcoxon test being a distribution-free test statistic under H_0, has all significance levels reasonably close to $\alpha = 0.05$.

6.21
   a. The *t*-test achieves its greatest power under the normal distribution.
   b. The Wilcoxon has greatest power under distributions like the double exponential and Cauchy distributions.

48 Chapter 6: Inferences Comparing Two Population Central Values

6.23
  a. Minitab output below
```
Paired T for A - B

            N     Mean    StDev   SE Mean
A           8    46.78     6.76      2.39
B           8    46.77     7.08      2.50
Difference  8    0.000    2.753     0.973


95% CI for mean difference: (-2.302, 2.302)
T-Test of mean difference = 0 (vs ≠ 0): T-Value = 0.00   P-Value = 1.000
```
  There is not significant evidence that the average difference between the treatments differs from 0 based on the paired t-test.

  b. Sign Test
```
Sign test of median =  0.00000 versus ≠ 0.00000

     N  Below  Equal  Above       P  Median
A-B  8      3      1      4  1.0000  0.4000
```
  There is not significant evidence that the average difference between the treatments differs from 0 based on the sign test.

  c. The results agree.
  d. The only difference is that the paired t-test requires the sample differences follow a normal distribution. If the conditions of the paired t-test are met, the power is higher.

6.25
  a. Here is a scatterplot:

  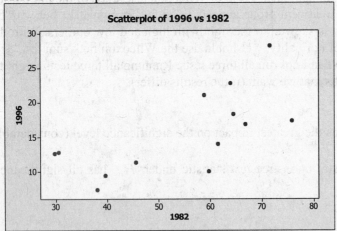

  There appears to be a positive correlation between the pairs of measurements.
  b. The correlation coefficient is $r = 0.671$. This confirms a positive relationship between the pairs of measurements.

c. $H_0: \mu_{1996-1982} \geq 0$ versus $H_a: \mu_{1996-1982} < 0$

$t = \dfrac{-38.80}{12.46/\sqrt{13}} = -11.23$, df = 12 $\Rightarrow$ p-value = $P(t \leq -11.23) < 0.0005$. We reject the null hypothesis and conclude that there has been a significant decrease in the mean PCB content of herring gull eggs.

$\mu_d \pm t \dfrac{s_d}{\sqrt{n}} = -38.80 \pm 2.179 \dfrac{12.46}{\sqrt{13}} = -38.80 \pm 7.53 = (-46.33, -31.27)$ We are 95% confident that the decrease in the mean PCB content of herring gull eggs is between 31.27 and 46.33.

We arrive at the same conclusion as we did in Exercise 6.11, but the magnitude of the test statistic is higher when the data are correctly treated as paired.

6.27
  a. To conduct the study using independent samples, the 30 participants should be very similar relative to age, body fat percentage, diet, and general health prior to the beginning of the study. The 30 participants would then be randomly assigned to the two treatments.
  b. The participants should be matched to the greatest extent possible based on age, body fat percentage, diet, and general health before the treatment is applied. Once the 15 pairs are configured, the two treatments are randomly assigned within each pair of participants.
  c. If there is a large difference in the participants with respect to age, body fat percentage, diet, and general health and if the pairing results in a strong positive correlation in the responses from paired participants, then the paired procedure would be more effective. If the participants are quite similar in the desired characteristics prior to the beginning of the study, then the independent samples procedure would yield a test statistic having twice as many degrees of freedom as the paired procedure and hence would be more powerful.

6.29
  a. $H_0: \mu_d = 0$ versus $H_a: \mu_d \neq 0$

  $t = 4.95$, df = 29 $\Rightarrow$ p-value = $2P(t \geq 4.95) < 0.001$ There is significant evidence of a difference in the mean final grades.
  b. A 95% confidence interval estimate of the mean difference in mean final grades is (2.23, 5.37).
  c. We would need to verify that the differences in the grades between the 30 twins are independent. The normal probability plot would indicate that the differences are a random sample from a normal distribution. Thus, the conditions for using a paired t-test appear to be valid.
  d. Yes. The purpose of pairing is to reduce the subject-to-subject variability, and there appears to be considerable differences in the students in the study. Also, a scatterplot of the data yields a strong positive correlation between the scores for the twins.

6.31
  a. Wilcoxon signed-rank test with $\alpha = 0.05$:
  ```
  Test of median = 0.000000 versus median ≠ 0.000000

           N for  Wilcoxon           Estimated
       N   Test   Statistic     P    Median
  A-B  8   7      16.0      0.800    0.4000
  ```
  Reject if $T \leq 5$; $T = 16$; Therefore, there is insufficient evidence to conclude the median difference differs from 0.

  b. The results are (unsurprisingly) the same. This data clearly indicates a mean/median of zero.

6.33
  a. The actual level of the t-test appears to be highly affected by heavy-tailed distributions (Cauchy).
  b. Heavy-tailness (Cauchy) has a greater effect of the level of the test (the level drops).
  c. No, the actual level of the Wilcoxon signed-rank test does not appear to be affected by sample size.
  d. The level of the Wilcoxon signed-rank test does not depend on the shape of the underlying distribution.

6.35
  a. The boxplot and normal probability plot both indicate that the distribution of the data is somewhat skewed to the left. Hence, the Wilcoxon signed-rank test would be more appropriate, although the paired t-test would not be inappropriate since the differences are nearly normal in distribution.
  b. H_0: The distribution of differences (female minus male) is symmetric about 0 versus H_a: The differences (female minus male) tend to be larger than 0.
  With $n = 20$, $\alpha = 0.05$, $T = T_-$, reject H_0 if $T_- \leq 60$.
  From the data, we obtain $T_- = 18 < 60$ and thus reject H_0 and conclude that repair costs are generally higher for female customers.

6.37 One-sided research hypothesis: $\mu_T > \mu_P$; $\sigma \approx 18.6$; $\alpha = 0.05$; $\beta \leq 0.20$ whenever $\mu_T - \mu_P > 5$; $n_T = n_P = n$

$$n \approx \frac{2\sigma^2(z_\alpha + z_\beta)^2}{\Delta^2} = \frac{2(18.6)^2(1.645 + 0.84)^2}{5^2} = 170.9 \Rightarrow n = 171$$

6.39
  a. $n = \dfrac{z_{\alpha/2}^2 \sigma_d^2}{E^2} = \dfrac{1.96^2(18.6)^2}{2.5^2} = 212.65 \Rightarrow n = 213$
  b. $n_P = \dfrac{(1.5)z_{\alpha/2}^2 \sigma_d^2}{E^2} = \dfrac{(1.5)1.96^2(18.6)^2}{2.5^2} = 318.97 \Rightarrow n_P = 319, n_T = 638$

6.41 $n = \dfrac{z_{\alpha/2}^2 \sigma_d^2}{E^2} = \dfrac{2.58^2 20^2}{4.25^2} = 147.41 \Rightarrow n = 148$

6.43
  a. $H_0: \mu_{\text{Narrow}} = \mu_{\text{Wide}}$ versus $H_a: \mu_{\text{Narrow}} \neq \mu_{\text{Wide}}$; $t = \dfrac{118.37 - 110.20}{\sqrt{(7.87)^2/12 + (4.71)^2/15}} = 3.17$, df $\approx 17 \Rightarrow$

  $0.002 <$ p-value $< 0.010 \Rightarrow$ Reject H_0 and conclude that there is sufficient evidence in the data that the two types of jets have different average noise levels.
  b. A 95% CI on $\mu_{\text{Narrow}} - \mu_{\text{Wide}}$ is $(2.73, 13.60)$
  c. Because maintenance could affect noise levels, jets of both types from several different airlines and manufacturers should be selected. They should be of approximately the same age, etc. This study could possibly be improved by pairing Narrow and Wide body airplanes based on factors that may affect noise level.

6.45 The problem is that the effect of soil characteristics, drainage, wind shelter, etc. on parasite counts could mask or exaggerate the effect of the fumigants on parasite counts.

6.47
a. $H_0: \mu_{Within} = \mu_{Out}$ versus $H_a: \mu_{Within} \neq \mu_{Out}$;

Since both n_1 and n_2 are greater than 10, the normal approximation can be used. $T = 122$, $\mu_T = (12)(12+14+1)/2 = 162$, $\sigma = \sqrt{(12)(14)(12+14+1)/12} = 19.44$

$z = \dfrac{122-162}{19.44} = -2.06 \Rightarrow$ p-value = 0.0394 $\Rightarrow$ Reject H_0 and conclude that the data provides sufficient evidence that there is a difference in average population abundance.

b. The Wilcoxon rank sum test requires independently selected random samples from two populations which have the same shape but may be shifted from one another.

c. The two populations may have very different variances, but the Wilcoxon rank sum test is very robust to departures from the required conditions.

d. The separate-variance test failed to reject H_0 with a p-value of 0.384. The Wilcoxon test rejected H_0 with a p-value of 0.0394. The difference in the two procedures is probably due to the skewness observed in the Outside data set. This can result in inflated p-values for the *t*-test, which relies on a normal distribution when the sample sizes are small.

6.49
a. Such a statement cannot be made. In order to study the effect of the oil spill on the population, we would need some baseline data at these sites before the oil spill.

b. If baseline data were available, a paired *t*-test would be appropriate. The impact of the oil spill could be studied by a *before* and *after* analysis. If the pairing were effective, the paired data study would be more efficient than a two independent samples stud, i.e., it would take fewer observations (sites) to achieve the same level of precision.

c. Spatial correlation might be a problem. Factors such as weather, food supply, etc., which are outside the control of the researcher might mask or exaggerate the effect of the oil spill. These types of factors also prevent the researcher from making definitive *cause* and *effect* statements.

6.51
a. $H_0: \mu_{Low-dose} = \mu_{Control}$ versus $H_a: \mu_{Low-dose} \neq \mu_{Control}$

Separate-variance *t*-test: $t = -2.09$ with df ≈ 35, p-value = 0.044 $\Rightarrow$ Reject H_0 and conclude there is significant evidence of a difference in the mean drop in blood pressure between the low-dose and control groups.

b. A 95% CI on $\mu_{Low-dose} - \mu_{Control}$ is (−51.3, −0.8), i.e., the low-dose group's mean drop in blood pressure was, with 95% confidence, 0.8 to 51.3 points less than the mean drop observed in the control group.

c. Provided the researcher independently selected the two random samples of participants, the conditions for using a separate-variance *t*-test were satisfied since the plots do not detect a departure from a normal distribution, but the sample variances are somewhat different (1.7 to 1 ratio).

52   Chapter 6: Inferences Comparing Two Population Central Values

6.53
  a. All hypertensive rats from the specified strain.
  b. $n = \dfrac{2\sigma^2(z_\alpha + z_\beta)^2}{\Delta^2} = \dfrac{2(30)^2(1.645+1.645)^2}{10^2} = 194.834 \Rightarrow n = 195$
  c. $n = \dfrac{z_{\alpha/2}^2 \sigma_d^2}{E^2} = \dfrac{2.58^2 30^2}{2.5^2} = 958.522 \Rightarrow n = 959$

6.55
  a. A 95% CI on $\mu_F - \mu_M$ is $(-142.3, -69.1)$ thousands of dollars.
  b. Since $s_1 \approx s_2$, use the pooled t-test: $t = \dfrac{245.3 - 350.1}{57.2\sqrt{1/20 + 1/20}} = -5.85$ with df = 38, p-value < 0.0001 $\Rightarrow$ Reject H_0 and conclude that there is significant evidence that the mean campaign expenditures for females is less than the mean campaign expenditures for males.
  c. Yes, the difference could be as much as $142,300.
  d. The required conditions are that the two samples are independently selected from populations having normal distributions with equal variances. The boxplots do not reveal any indication that the population distributions were not normal. The sample variances have a ratio of 1.4 to 1.0, thus there is very little indication that the population variances were unequal.

6.57 Let d = Before – After
  a. $H_0: \mu_{\text{Before}} = \mu_{\text{After}}$ versus $H_a: \mu_{\text{Before}} \neq \mu_{\text{After}}$;
     $t = \dfrac{-0.122}{0.106/\sqrt{15}} = -4.45$ with df = 14, p-value < 0.0005 $\Rightarrow$ Reject H_0 and conclude that there is significant evidence that the mean soil pH has changed after mining on the land.
  b. $H_a: \mu_{\text{Before}} \neq \mu_{\text{After}}$
  c. A 99% CI on $\mu_{\text{Before}} - \mu_{\text{After}}$ is $(0.04, 0.20)$.
  d. The findings are highly significant (p-value < 0.0005), statistically. The question is, how significant are the results in a practical sense? Unless a change in pH of between 0.04 and 0.20 has an impact on the soil with respect to common usages of the soil, the mining company should not be cited.

6.59
  a. The average potency after one year is different than the average potency right after production.
  b. $t = t' = 4.2368$; The two test statistics are equal since the sample sizes are equal.
  c. p-value = 0.0006 for t; p-value = 0.0005 for t'; The p-values are different since the test statistics have different degrees of freedom.
  d. In this particular experiment, the test statistics reach the same conclusion, reject H_0.
  e. Since $s_1 \approx s_2$, which yields a test of equal variances with p-value = 0.3917, the pooled t-test would be the more appropriate test statistic.

6.61 A 95% CI on the difference in the mean flare-illumination values, $\mu_1 - \mu_2$:

$(190 - 213.4) \pm 2.101(12.64)\sqrt{\dfrac{1}{10} + \dfrac{1}{10}} = -23.4 \pm 11.88 = (-35.28, -11.52)$

6.63

a. Before running the tests, we must aggregate the number of seizures over the 8weeks by adding up y1 to y4. This is our response variable.

$H_0 : \mu_{Placebo} \leq \mu_{Progabide}$ versus $H_a : \mu_{Placebo} > \mu_{Progabide}$.

The p-value from the pooled *t*-test is 0.416 (from Minitab), which yields a decision of failure to reject H_0. Thus, there is insufficient evidence that Progabide reduces the mean number of seizures for epileptics.

Using Minitab, the Wilcoxon rank sum test yields:
```
95.1 Percent CI for η1 - η2 is (-3.99,12.01)
W = 925.5
Test of η1 = η2 vs η1 < η2

Cannot reject since W is > 840.0
```
$\Rightarrow$ Fail to reject H_0. Thus, the same conclusion is reached as with the *t*-test.

b. The boxplots indicate that both distributions may be skewed to the right. Thus, the Wilcoxon rank sum test may be more appropriate than the *t*-test. However, the sample sizes are relatively large, which would indicate the *t*-test would not be totally inappropriate.

c. A 95% CI on the difference in the mean number of seizures, $\mu_{Placebo} - \mu_{Progabide}$:

(-21.4, 26.6)—see below.

Two-Sample T-Test and CI: Seiz, Trt

```
Two-sample T for Seiz

Trt   N   Mean   StDev   SE Mean
0    28   34.4   35.1    6.6
1    31   31.8   53.9    9.7


Difference = μ (0) - μ (1)
Estimate for difference: 2.6
95% CI for difference: (-21.4, 26.6)
T-Test of difference = 0 (vs ≠): T-Value = 0.21   P-Value = 0.832   DF = 57
Both use Pooled StDev = 45.9639
```

6.65

a. $n = \dfrac{2(700)^2 (1.96)^2}{(500)^2} = 15.1 \Rightarrow n = 16$. The new study requires 16 vehicles of each type.

b. $n = \dfrac{2(700)^2 (1.645 + 1.645)^2}{(500)^2} = 42.3 \Rightarrow n = 43$. The new study requires 43 vehicles of each type.

6.67

a. The following table contains statistics for the three variables

	Mean	SD	Minimum	Q1	Median	Q3	Maximum
Exposed	31.85	14.41	10.00	20.50	34.00	40.00	73.00
Control	15.88	4.54	7.00	12.50	16.00	19.00	25.00
Difference	15.97	15.86	−9.00	3.00	15.00	25.00	60.00

i. The median of the Exposed children (34) is large than the maximum (25) of the Controls. This confirms the statement "We can see than over half of the children in the Exposed group have more lead in their blood than do any of the children in the Control group" (p. 356).
ii. The difference between the individual means is equal to the mean of the differences (15.97). On average, the lead values for the Exposed children are 15.97 micrograms per deciliter of blood higher than the lead values for the Control children.

b. The median is not the same because the median of the Difference is the median of a new set of values (the differences) of the 33 pairs of children. If you wanted to give the most accurate view of the increase in lead exposure due to a parent working at the battery factory, you would use the median of the differences: 15 micrograms per deciliter of blood. The differences take into account the pairing of the children in the study which helps to reduce the impact of factors other than employment by the parents in the battery factory.

6.69

a. The normal probability plot for the type V subjects appears to be fairly normal, but the normal probability plot for the normal subjects appears to be highly skewed (likely due to the large outlier).

b. Pooled t-test

Two-Sample T-Test and CI: Type V, Normal

```
Two-sample T for Type V vs Normal

         N   Mean   StDev   SE Mean
Type V   9   2.14   1.58    0.53
Normal   9   100    132     44


Difference = μ (Type V) - μ (Normal)
Estimate for difference:  -97.9
99% CI for difference:  (-226.1, 30.4)
T-Test of difference = 0 (vs ≠): T-Value = -2.23  P-Value = 0.040  DF = 16
Both use Pooled StDev = 93.1427
```

Separate Variance t-test

Two-Sample T-Test and CI: Type V, Normal

```
Two-sample T for Type V vs Normal

         N   Mean   StDev   SE Mean
Type V   9   2.14   1.58    0.53
Normal   9   100    132     44


Difference = μ (Type V) - μ (Normal)
Estimate for difference:  -97.9
99% CI for difference:  (-245.2, 49.5)
T-Test of difference = 0 (vs ≠): T-Value = -2.23  P-Value = 0.056  DF = 8
```

Wilcoxon Signed Rank
Mann-Whitney Test and CI: Type V, Normal

```
         N   Median
Type V   9    2.0
Normal   9   44.8


Point estimate for η1 - η2 is -42.8
99.2 Percent CI for η1 - η2 is (-134.6,-22.5)
W = 45.0
Test of η1 = η2 vs η1 ≠ η2 is significant at 0.0004
The test is significant at 0.0004 (adjusted for ties)
```

c. Part a suggests the normality assumption is not satisfied which means the signed rank (nonparametric) test is advised. Both the pooled *t*-test and the separate-variance *t*-test would fail to reject H_0 at $\alpha = 0.01$ significance level, but the (correct) Wilcoxon rank sum test will reject H_0 at $\alpha = 0.01$.

Chapter 7

Inferences about Population Variances

7.1
- a. Cultures of *E. coli* tested by the Petrifilm HEC test and cultures of *E. coli* tested by HGMF.
- b. The concentration of *E. coli* in the culture, measurement error, human error.
- c. We have 48 samples and number them 1, 2,..., 48. Use a computer program to produce 24 randomly selected numbers between 1 and 48. The samples having these values are assigned to HEC, and the other 24 samples are assigned to HGMF.
- d. Do HEC and HGMF have the same degree of variability in their determination of *E. coli* level? Do the HEC and HGMF procedures yield the same mean *E. coli* concentrations when applied to the same samples? Are the proportion of positive tests the same for both procedures?

7.3
- a. 28.87
- b. 34.81
- c. 9.390
- d. 31.53
- e. 30.8
- f. 9.74

7.5
- a. Let y be the quantity in a randomly selected jar:
  $P(underfilled) = P(y < 64.3) = P\left(z < \frac{64-64.3}{0.15}\right) = P(z < -2) = 0.0228 \Rightarrow 2.28\%$
  $P(overfilled) = 1 - P(underfilled) = 1 - 0.0228 = 0.9772 \Rightarrow 97.72\%$

- b. $s = 0.1893$

  95% CI for σ $\left(\sqrt{\frac{(24-1)(0.1893)^2}{38.08}}, \sqrt{\frac{(24-1)(0.1893)^2}{11.69}}\right) \Rightarrow (0.1471, 0.2655)$

  Because the 95% CI includes 0.15, the results are consistent with a process standard deviation of 0.15.

- c. $H_0: \sigma \leq 0.15$ versus $H_A: \sigma > 0.15$

  Reject H_0 if $\frac{(n-1)s^2}{0.15^2} > 35.17$; $\frac{(24-1)(0.1893)^2}{0.15^2} = 36.63 > 35.17$

  Reject H_0 and conclude the data supports σ greater than 0.15.

  Note: This is in contradiction to part c because part c was a 2-sided interval and the test in part d was one-sided.

- d. $p - value = P\left(\frac{(n-1)s^2}{0.15^2} \geq 36.63\right) = 0.0355$ (from software)
- e. The assumption of normality appears satisfied here so the test is valid.
- f. All containers produced by this process at this facility.

7.7
- a. The boxplot is symmetric, but there are four outliers. Since the sample size is 150, a few outliers would be expected. However, 4 of the 150 may indicate the population may have heavier tails than a normal distribution. This may cause the value of s to be inflated.
- b. 99% CI on σ: $\left(\sqrt{\frac{(150-1)(9.537)^2}{197.21}}, \sqrt{\frac{(150-1)(9.537)^2}{108.29}}\right) \Rightarrow (8.290, 11.187)$
- c. $H_0: \sigma^2 \leq 9$ vs. $H_A: \sigma^2 > 9$
  Reject H_0 if $\frac{(n-1)s^2}{9^2} \geq 192.073$; $\frac{(150-1)9.537^2}{9^2} = 167.3111 < 192.073 \Rightarrow$
  Fail to reject H_0 and conclude the data fail to support the statement that σ^2 is greater than 9.

7.9
- a. The boxplot is symmetric with no outliers.
- b. 95% CI on σ: $\left(\sqrt{\frac{(40-1)(2.684)^2}{19.996}}, \sqrt{\frac{(40-1)(2.684)^2}{65.476}}\right) \Rightarrow (2.071, 3.748)$
- c. Based on the Minitab output below, there is insufficient evidence to conclude the mean rebound coefficient is less than 85 at the 0.05 level.

One-Sample T: Rebound

```
Test of μ = 85 vs < 85

Variable   N    Mean   StDev   SE Mean   95% Upper Bound      T      P
Rebound    40   84.797  2.684   0.424              85.513   -0.48  0.318
```

- d. $H_0: \sigma^2 \leq 2$ vs. $H_A: \sigma^2 > 2$
  Reject H_0 if $\frac{(n-1)s^2}{2^2} \geq 62.428$; $\frac{(40-1)2.684^2}{2^2} = 70.2376 < 62.428 \Rightarrow$
  Reject H_0 and conclude the data support the statement that σ^2 is greater than 2.
- e. We can apply this inference to all baseballs from the supplier(s) that supplied that purchaser.

7.11
- a. $F_{0.05,7,9} = 3.2927$
- b. $F_{0.025,9,7} = 4.8232$
- c. $F_{0.01,17,9} = 4.8902$
- d. $F_{0.1,9,20} = 1.9649$
- e. $F_{0.25,15,12} = 1.4796$
- f. $F_{0.15,15,19} = 1.6518$

7.13
- a. $F_{0.05,14,9} = 3.0255$
- b. $F_{0.025,39,27} = 2.0744$
- c. $F_{0.01,50,39} = 2.0723$
- d. $F_{0.1,39,40} = 1.5085$
- e. $F_{0.001,45,45} = 2.5677$
- f. $F_{0.005,25,39} = 2.4991$

7.15

a. 95% CI on σ_{Old}: $\left(\sqrt{\dfrac{(61-1)(0.231)^2}{83.30}}, \sqrt{\dfrac{(61-1)(0.231)^2}{40.48}} \right) \Rightarrow (0.196, 0.281)$

95% CI on σ_{New}: $\left(\sqrt{\dfrac{(61-1)(0.162)^2}{83.30}}, \sqrt{\dfrac{(61-1)(0.162)^2}{40.48}} \right) \Rightarrow (0.137, 0.197)$

b. $H_0: \sigma_{Old}^2 \leq \sigma_{New}^2$ versus $H_a: \sigma_{Old}^2 > \sigma_{New}^2$

With $\alpha = 0.05$, reject H_0 if $\dfrac{s_{Old}^2}{s_{New}^2} \geq 1.53$

$\dfrac{s_{Old}^2}{s_{New}^2} = 2.033 > 1.53$

Thus, we reject H_0 and conclude the data supports the statement that σ_{New}^2 is less than σ_{Old}^2.

c. The boxplots indicate that both distributions are normally distributed. From the problem description, the two samples appear to be independently selected random samples.

7.17

a. While both heavy-tailedness and skewness have an effect on the F-test, skewness (Gamma distributions) has more of an impact than does heavy-tailedness.

b. Increasing the sample size for the normal and uniform distributions tends to decrease the level of the F-test. This makes sense, given that the variation of the statistics decreases as the sample sizes increase. For the t and Gamma distributions (heavy-tailed and skewed, respectively), the level of the F-test increases with an increase in both sample sizes. One reason for this happening is that for heavy-tailed and highly skewed distributions, larger sample sizes increase the chance of obtaining extreme values, which yields values of the sample standard deviations (and hence test statistic) different from what would be expected under sampling from a normal population.

c. A negative impact is that the actual level of the test has been decreased from the specified value of $\alpha = 0.05$, and thus the power of the test will be smaller than what would be expected under the nominal level of the test. This will result in an inflated chance of a Type II error occurring. A positive impact is that the actual level of the test has been decreased, and thus the chance of a Type I error has decreased as well.

7.19 The skewness in the data produces outliers that may greatly distort both the mean and standard deviation. Thus, the BFL test statistic minimizes both of these effects by replacing the mean with the median and using the absolute deviations about the median in place of the squared deviations about the mean.

7.21
   a. $25 \times 90\% = 22.5$ and $25 \times 110\% = 27.5$ implies the limits are 22.5 and 27.5
   b. The boxplot and normal probability plot are given here:

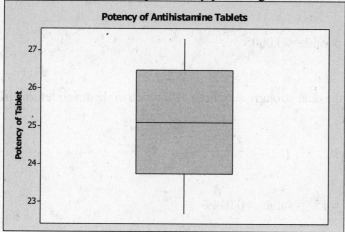

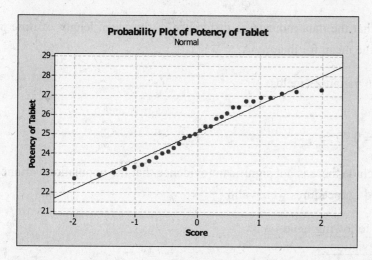

The boxplot indicates a symmetric distribution with no outliers. The normal probability plot shows the data values reasonably close to a straight line, although there is some deviation at both ends, which indicates that the data may be a random sample from a distribution which has shorter tails than a normally distributed population.

   c. Range $= 27.5 - 22.5 = 5 \Rightarrow \hat{\sigma} = 5/4 = 1.25$
   $H_0: \sigma = 1.25$ versus $H_a: \sigma \neq 1.25$

   With $\alpha = 0.05$, reject H_0 if $\dfrac{(n-1)s^2}{(1.25)^2} \leq 16.05$ or $\dfrac{(n-1)s^2}{(1.25)^2} \geq 45.72$

   $\dfrac{(30-1)(1.4691)^2}{(1.25)^2} = 40.06 \Rightarrow 16.06 < 40.06 < 45.72$

   Thus, we fail to reject H_0 and conclude there is insufficient evidence that the product standard deviation is greater than 1.25. Thus, it appears that the potencies are within the required bounds.

7.23 The boxplots indicate that data from both portfolios have a normal distribution. Also, the CI on the ratio of the variances contained 1, which indicates equal variances. Thus, a pooled variance t-test will be used as the test statistic.

$H_0: \mu_1 = \mu_2$ versus $H_a: \mu_1 \neq \mu_2$

$t = \dfrac{131.60 - 147.20}{4.92\sqrt{\dfrac{1}{10} + \dfrac{1}{10}}} = -7.09$, df = 18, p-value < 0.0005

Thus we reject H_0 and conclude that the data strongly support a difference in the mean returns of the two portfolios.

7.25 We would now run a one-tailed test:

$H_0: \mu_A \geq \mu_B$ versus $H_a: \mu_A < \mu_B$

$t = \dfrac{27.62 - 34.69}{\sqrt{\dfrac{(9.83)^2}{13} + \dfrac{(4.03)^2}{13}}} = -2.40 \Rightarrow df = 15$, p-value $= 0.0149$

Thus we reject H_0 and conclude that the data indicates a difference in the mean length of time people remain on therapy B is longer than the mean for therapy A.

7.27 A comparison of the population variances yields:

$H_0: \sigma_D^2 = \sigma_I^2$ versus $H_a: \sigma_D^2 \neq \sigma_I^2$

$\dfrac{s_D^2}{s_I^2} = \dfrac{(2.956)^2}{(2.565)^2} = 1.328 \Rightarrow$ p-value $= 0.70 \Rightarrow$

Thus, we fail to reject H_0 and conclude there is not significant evidence that the population variances are different. This means that the pooled t procedures were valid.

7.29 The data is summarized in the following table:

Oven	n	Mean	Standard Deviation
1	10	1670.81	1.27737
2	10	1670.35	1.50508
3	10	1670.12	6.53761

a. Boxplots are given here:

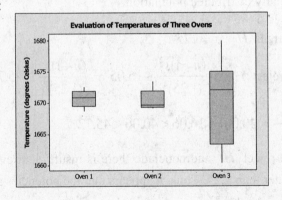

From the boxplots, we observe that the data from Oven 1 appears to be from a normal distribution since the boxplot has a symmetric box with whiskers of approximately equal length and no outliers. The boxplot for the data for Oven 2 is skewed to the right, while the boxplot for the data for Oven 3 is very slightly skewed to the left. Thus, using the BFL test rather than the Hartley test was the appropriate choice.

Test $H_0: \sigma_1 = \sigma_2 = \sigma_3$ versus H_a: population standard deviations are not all equal

Reject H_0 at level $\alpha = 0.05$ if $L \geq F_{0.05, 2, 27} = 3.35$

From the data, $L = 7.97 > 3.35 \Rightarrow$

Thus we reject H_0 and conclude there is a significant evidence of a difference in variability of the temperature of the ovens.

b. 95% CI on $\dfrac{\sigma_1}{\sigma_2}$: $\left(\sqrt{\dfrac{(1.27737)^2}{(1.50508)^2}(0.248)}, \sqrt{\dfrac{(1.27737)^2}{(1.50508)^2}(4.03)} \right) \Rightarrow (0.42, 1.70)$

95% CI on $\dfrac{\sigma_1}{\sigma_3}$: $\left(\sqrt{\dfrac{(1.27737)^2}{(6.53761)^2}(0.248)}, \sqrt{\dfrac{(1.27737)^2}{(6.53761)^2}(4.03)} \right) \Rightarrow (0.10, 0.39)$

95% CI on $\dfrac{\sigma_2}{\sigma_3}$: $\left(\sqrt{\dfrac{(1.50508)^2}{(6.53761)^2}(0.248)}, \sqrt{\dfrac{(1.50508)^2}{(6.53761)^2}(4.03)} \right) \Rightarrow (0.11, 0.46)$

c. We used the BFL test in part (a) because of the skewed distributions of the data for Oven 2. Because one of the distributions was skewed (not normal), the confidence intervals we formed may be risky estimations.

7.31
a. The boxplots show that both distributions are skewed right. The variability of the two distributions do not appear to be different. Thus, we will use the pooled t-test.

$H_0: \mu_U \geq \mu_S$ versus $H_a: \mu_U < \mu_S$

$t = 2.00$, df = 50, p-value = 0.026

Thus we reject H_0 and conclude that the data strongly support that seeding has increased the mean rainfall.

b. Test $H_0: \sigma_U = \sigma_S$ versus $H_a: \sigma_U \neq \sigma_S$

Because of the right skewness of the data, the BFL will be used. $L = 2.86$ with p-value = 0.097.

Thus, we fail to reject H_0 and conclude there is not significant evidence of a difference in the level of variability in the amount of rainfall between seeded and unseeded clouds.

c. $H_0: \mu_S - \mu_U \leq 100$ versus $H_a: \mu_S - \mu_U > 100$

$t = \dfrac{(442 - 164.5) - 100}{500.529 \sqrt{\dfrac{1}{26} + \dfrac{1}{26}}} = 1.278$, df = 50, p-value = 0.104

Thus we fail to reject H_0 and conclude that the data do not support that seeding is economically viable.

7.33
  a. All urban households of the given sizes.
  b. Based on the BFL-test's p-value of 0.44 (see below), we would fail to reject the hypothesis of equal variances. See Minitab output below. The level of variation is not significantly different among the groups.

Test for Equal Variances: Expendit versus NumberMe

```
Method

Null hypothesis         All variances are equal
Alternative hypothesis  At least one variance is different
Significance level      α = 0.05


95% Bonferroni Confidence Intervals for Standard Deviations

NumberMe   N    StDev         CI
       1  20  37.9915  (22.7433,  72.845)
       2  23  20.1869  (14.8407,  30.922)
       3  16  33.6248  (18.1342,  74.311)
       4  14  44.6505  (16.9260, 144.345)
       5  10  28.7999  ( 7.6725, 145.611)

Individual confidence level = 99%




Tests

                         Test
Method                Statistic  P-Value
Multiple comparisons      —        0.247
Levene                   0.95      0.440
```

Chapter 8

Inferences about More Than Two Population Central Values

8.1
   a. The populations of interest are people with port-wine stains who are in the 0-5 year, 6-11 year, 12-17 year, and 18-31 year age groups.
   b. Other factors of interest might be size and location of the port-wine stain.

8.3
   a. Third grade students in this district in comparable schools.
   b. Third grade students in this district in comparable schools.
   c. The analysis of variance procedures would be okay assuming that we satisfy the conditions that each of classes has a score distribution that is normal, that the variances of the score distribution are the same across all classes, and that we have independent random samples. The question here is whether students are randomly assigned to each of the classes. If, for example, students with better grades have first choice of teacher and time of class, there might be a systematic bias in the difference in means before the study is done. It is also important to take into account the differences in the five schools. This will be discussed in Chapter 15.

8.5
   a. We have that $n_1 = n_2 = n_3 = n_4$. Let the common sample size be denoted n_C. In that case,
   $$s_W^2 = \frac{SSW}{n_T - 4} = \frac{\sum(n_i - 1)s_i^2}{4n_C - 4} = \frac{(n_C - 1)\sum s_i^2}{4(n_C - 1)} = \frac{\sum s_i^2}{4}$$
   b. This relationship does not hold if the sample sizes are not equal. The reason that we shouldn't just take an average when the sample sizes are not equal is that we would be giving the same weight to the variances of samples with potentially disparate sample sizes (suppose one sample is size 5 and another is size 20). We should be weighting the variances appropriately, and a simple average does not do this when the sample sizes differ.

8.7
   a. Based on the ANOVA table below, there is not a significance difference in soil density among the grazing regimens at the $\alpha = 0.05$ level.
   ```
   Analysis of Variance

   Source  DF  Adj SS  Adj MS  F-Value  P-Value
   Factor   2   2.692  1.3460    2.06    0.156
   Error   18  11.738  0.6521
   Total   20  14.430
   ```
   b. The p-value is 0.156.
   c. The standard deviation of the two-week grazing, two-week no grazing group is smaller (about half) of the other two. This may call into question our assumption of equal variance, but with the small sample size, it's likely ok to proceed.

8.9 $y_{ij} = \mu + \tau_i + \varepsilon_{ij}$

a. $t = 3$; $n_1 = n_2 = n_3 = 5$

b. $\mu = 11$; $\tau_1 = -1.8$; $\tau_2 = -1$; $\tau_3 = 2.8$; $\sigma = \sqrt{\frac{4(33.7)+4(29.0)+4(46.7)}{12}} = 6.04$

c. $\mu = 13.8$; $\tau_1 = -4.6$; $\tau_2 = -3.8$; $\tau_3 = 0$; $\sigma = 6.04$

d. The difference is that part b uses the overall average (grand mean) as the baseline and compares all the treatments to it. Part b uses variety 3 as the baseline to which other varieties are compared.

8.11

a. If $s_W^2 = 0$, there is no variability in the groups. This means that the groups must be significantly different is any of the sample means differ from another.

b. Because within variability is 0, every point in a given class must lie at the class mean. Therefore, the residuals are all 0.

8.13

a. $s_W^2 = \frac{8(5.625)^2 + 7(3.101)^2 + 9(3.674)^2}{24} = 18.41$ whereas the average of the variances is 18.25

b. A boxplot of test scores by method is give here:

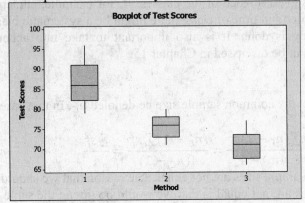

Since the sample sizes are small, it is difficult to judge normality or homogeneity of variance in this case. Based on a cursory overview, there do not appear to be any violations of AOV assumptions but a larger sample size is desired.

A normal probability plot and boxplot of the residuals are given here:

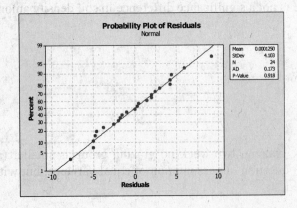

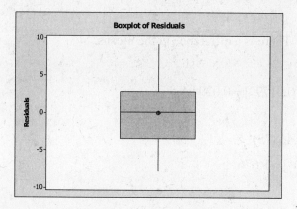

Both the normal probability plot and the boxplot show that the data are approximately normal. The test for lack of normality has a p-value of 0.918, so we are willing to assume normality in this situation. The BFL test yields $L = 1.68$ with p-value = 0.210. There is not sufficient evidence of a difference in the variances of the test scores for the three methods. AOV procedures appear to be satisfied.

8.15

   a. The probability plots do not look very normal. Tests of normality also indicate deviations from normality.

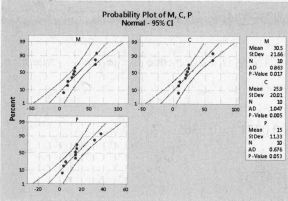

   b. The transformation of taking the absolute value of the difference of score with the group median doesn't appear to improve the normality of the data a great deal if at all.

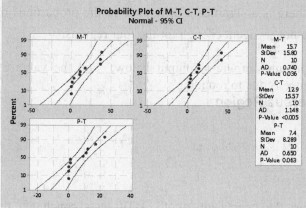

8.17

a. $H_0: \mu_{NE} = \mu_{SE} = \mu_{MW} = \mu_W$ versus H_a: There is a difference in the means.

Reject H_0 if $F \geq F_{0.05,3,20} = 3.10$

$SSW = 5[0.0273^2 + 0.0638^2 + 0.0274^2 + 0.0179^2] = 0.0294$

$\bar{y}_{..} = 0.46875 \Rightarrow$

$SSB = 6[(0.827 - 0.46875)^2 + (0.343 - 0.46875)^2$

$+ (0.585 - 0.46875)^2 + (0.120 - 0.46875)^2] = 1.676$

$F = \dfrac{1.676/3}{0.0294/20} = 380.05 > 3.10 \Rightarrow$

Thus, we reject H_0 and conclude there is a significant difference in the proportions of people who thought the EPA standards were not stringent enough for the four regions.

The data was analyzed using a computer program (output below), and the value of F was determined to be 379.34. The difference is due to rounding.

One-way ANOVA: Proportion versus Region

```
Source  DF      SS       MS        F       P
Region   3  1.67385  0.55795  379.34  0.000
Error   20  0.02942  0.00147
Total   23  1.70326
```

b. $H_0: \mu^*_{NE} = \mu^*_{SE} = \mu^*_{MW} = \mu^*_W$ versus H_a: There is a difference in the means, where μ^* are the means of the transformed data.

Reject H_0 if $F \geq F_{0.05,3,20} = 3.10$

$SSW = 5[0.0354^2 + 0.0673^2 + 0.0279^2 + 0.0271^2] = 0.0365$

$\bar{y}_{..} = 0.74775 \Rightarrow$

$SSB = 6[(1.142 - 0.74775)^2 + (0.625 - 0.74775)^2$

$+ (0.871 - 0.74775)^2 + (0.353 - 0.74775)^2] = 2.049$

$F = \dfrac{2.049/3}{0.0365/20} = 374.25 > 3.10 \Rightarrow$

Thus, we reject H_0 and conclude there is a significant difference in the proportions of people who thought the EPA standards were not stringent enough for the four regions.

The data was analyzed using a computer program (output below), and the value of F was determined to be 374.64. The difference is due to rounding.

One-way ANOVA: TransProp versus Region

```
Source  DF      SS       MS        F       P
Region   3  2.05044  0.68348  374.64  0.000
Error   20  0.03649  0.00182
Total   23  2.08692
```

c. Transforming the data did not alter the conclusion—both AOV tests concluded there is a significant difference in the proportions of people who thought the EPA standards were not stringent enough for the four regions.

8.19
a. The boxplots of the original weights by environment are given here:

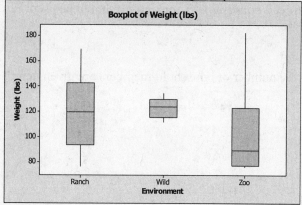

The weights for deer on the ranch appear symmetric, while the weights for deer in the wild are slightly skewed to the left and the weights for deer in the zoo are very skewed to the right.

A normal probability plot and boxplot of the residuals are given here:

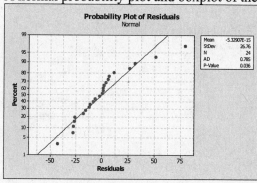

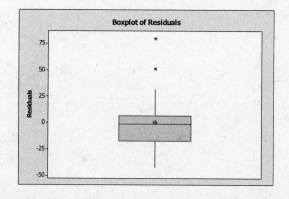

The test for lack of normality has a p-value of 0.036, so we should not be willing to assume normality in this situation. AOV procedures do not appear to be satisfied.

However, the BFL test for equality of variances gives 1.44 with a p-value of 0.260. It appears that the three populations have variances that are approximately equal.

68   Chapter 8: Inferences about More Than Two Population Central Values

b. No transformation of the data will fix the skewness of the zoo weights while still keeping the wild and ranch weights fairly normal. The BFL test failed to reject the hypothesis that the three population variances are equal. Therefore, a nonparametric procedure should be used. See Section 8.6.

8.21
a. The boxplots of the food expenditures by number of household members are given here:

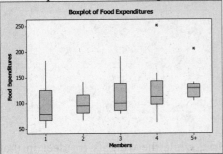

The food expenditures appear skewed right for households with 1, 2, 3, or 4 members. For 5+ member households, the food expenditures appear skewed left, and there is a high outlier.

A normal probability plot and boxplot of the residuals are given here:

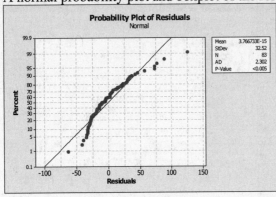

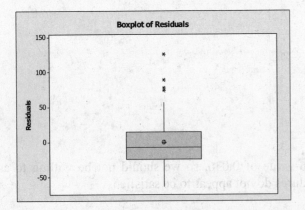

The test for lack of normality has a p-value < 0.005, so we should not be willing to assume normality in this situation. The boxplot of residuals confirms this, as the residuals are highly right skewed. AOV procedures do not appear to be satisfied.

b. The log transformation yields a normal test p-value of 0.062, thus we will take the logarithms of the data.

$H_0: \mu_1^* = \mu_2^* = \mu_3^* = \mu_4^* = \mu_5^*$ versus H_a: There is a difference in the means, where μ^* are the means of the logarithms of the data.

Reject H_0 if $F \geq F_{0.05, 4, 78} = 2.49$

One-way ANOVA: log Expend versus Members
```
Source   DF      SS      MS      F      P
Members   4  1.4788  0.3697   4.54  0.002
Error    78  6.3567  0.0815
Total    82  7.8354
```
From the Minitab output, we have that F = 4.54 > 2.49 (p-value = 0.002).

Thus, we reject H_0 and conclude there is a significant difference in the mean food expenditures for the different household sizes.

8.23
a. The boxplots of reliability by plant are given here:

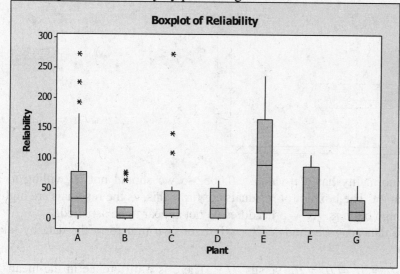

All of the distributions appear to be right skewed.

A normal probability plot and boxplot of the residuals are given here:

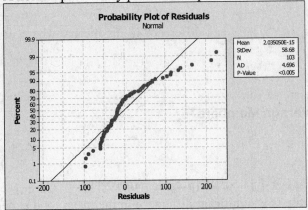

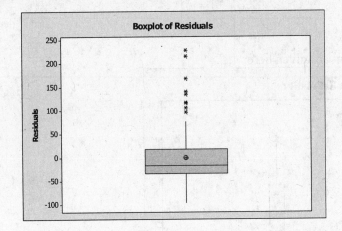

The test for lack of normality has a p-value < 0.005, so we should not be willing to assume normality in this situation. The boxplot of residuals confirms this, as the residuals are highly right skewed with several high outliers. AOV procedures do not appear to be satisfied.

b. Since we are told the data may be modeled by a Poisson, we transform the data by taking the square root.

$H_0: \mu_A^* = \mu_B^* = \mu_C^* = \mu_D^* = \mu_E^* = \mu_F^* = \mu_G^*$ versus H_a: There is a difference in the means, where μ^*'s are means of the transformed data.

Reject H_0 if $F \geq F_{0.05, 6, 96} = 2.19$

One-way ANOVA: sqrtReliability versus Plant
```
Source   DF      SS      MS      F       P
Plant     6   252.5    42.1   2.68   0.019
Error    96  1506.5    15.7
Total   102  1759.0
```

From the Minitab output, we have that F = 2.68 > 2.19 (p-value = 0.019).

Thus, we reject H_0 and conclude there is a significant difference in the mean reliability among the plants.

c. H_0: There is no difference among the seven plants with respect to reliability.

H_a: At least one of the seven plants differs from the others with respect to reliability.

The Minitab output for the Kruskal-Wallis test is given here:
Kruskal-Wallis Test: Reliability versus Plant
```
Kruskal-Wallis Test on Reliability

Plant       N    Median    Ave Rank      Z
A          34    34.000       56.6     1.10
B          15     6.000       35.4    -2.32
C          17    17.000       52.6     0.08
D           8    18.500       44.9    -0.70
E          12    90.500       70.5     2.28
F           7    18.000       58.1     0.56
G          10    14.500       39.5    -1.40
Overall   103                 52.0

H = 12.53   DF = 6   P = 0.051
H = 12.54   DF = 6   P = 0.051   (adjusted for ties)
```

At the 0.05 significance level, we fail to reject the null hypothesis since the p-value = 0.051 for the Kruskal-Wallis test. Thus, we fail to conclude that at least one of the seven plants differs from the others with respect to reliability.

d. We cannot use AOV assuming normality, so I would not trust that test. The Poisson transformation is good, and if the data can really be modeled by Poisson distributions, then we should use that result. However, since we cannot be sure, I am most confident about the Kruskal-Wallis test since it has fewer restrictions. At the same time, the p-value is so close to 0.051 that we need to ask ourselves if failing to reject with the Kruskal-Wallis test is meaningful.

8.25

a. H_0: There is no difference in among the three analgesics with respect to pain reduction.

H_a: At least one of the three analgesics differs from the others with respect to pain reduction.

The Minitab output for the Kruskal-Wallis test is given here:
Kruskal-Wallis Test: Pain Reduction versus Analgesic
```
Kruskal-Wallis Test on Pain Reduction

Analgesic   N    Median    Ave Rank      Z
1          15     1.000       14.0    -3.25
2          15     3.700       25.1     0.77
3          15    10.000       29.9     2.48
Overall    45                 23.0

H = 11.54   DF = 2   P = 0.003
H = 11.55   DF = 2   P = 0.003   (adjusted for ties
```

At the 0.05 significance level, we reject the null hypothesis since the p-value = 0.003 for the Kruskal-Wallis test. Thus, we conclude that at least one of the three analgesics differs from the others with respect to pain reduction.

b. We rejected the null hypothesis in Exercise 8.22 both with the original data and with the transformation, with p-values of 0.0008 and 0.0009. The p-value for the H test was 0.003, which is somewhat larger but still yields a rejection of the null hypothesis.

8.27

a. Levene's test gives a p-value of 0.042. This is significant at the 0.05 level suggesting there is a difference in the variances of the groups. We should be cautious about applying AOV methods.

b. Based on the AOV, there is significant evidence that the mean deviations differ among the suppliers. (p-value<0.05)

```
Analysis of Variance

Source  DF  Adj SS   Adj MS   F-Value  P-Value
Factor   4   28024  7006.09    265.94    0.000
Error   40    1054    26.34
Total   44   29078
```

c. The Kruskal-Wallis test agrees that there is a difference in the mean deviation among the suppliers. (p-value<0.05)

Kruskal-Wallis Test: Deviation versus Supplier

```
Kruskal-Wallis Test on Deviation

Supplier   N   Median   Ave Rank      Z
1          9    189.1       32.8    2.50
2          9    174.1       23.0    0.00
3          9    156.9       14.0   -2.30
4          9    132.9        5.0   -4.60
5          9    204.4       40.2    4.40
Overall   45                23.0

H = 41.59   DF = 4   P = 0.000
H = 41.60   DF = 4   P = 0.000   (adjusted for ties)
```

d. Based on the 95% confidence intervals for the means of the groups (seen below), the upper limit of the lowest deviation supplier (4) is more than 20 units below the lower limit of the highest deviation supplier (5). Therefore, we can conclude a practical difference.

```
Means

Supplier   N     Mean   StDev          95% CI
1          9  189.144   2.948   (185.687, 192.602)
2          9  174.000   2.624   (170.542, 177.458)
3          9   156.28    3.30   ( 152.82,  159.74)
4          9   132.61    5.00   ( 129.15,  136.07)
5          9   203.94    8.96   ( 200.49,  207.40)
```

8.29

a. Based on the boxplots and the normal probability plot, the condition of normality of the population distributions appears to be satisfied. The Hartley test yields:

$$F_{max} = \frac{(1.0670)^2}{(0.8452)^2} = 1.59 < 14.5$$

(value from Table 12 with $\alpha = 0.01$) $\Rightarrow$ There is not significant evidence of a difference in the four population variances.

b. From the ANOVA table, we have p-value < 0.001. Thus, there is significant evidence that the mean ratings differ for the four groups.

c. 95% CI on μ_I: $8.3125 \pm 2.048 \dfrac{\sqrt{0.9763}}{\sqrt{8}} \Rightarrow (7.6, 9.0)$

95% CI on μ_{II}: $6.4375 \pm 2.048 \dfrac{\sqrt{0.9763}}{\sqrt{8}} \Rightarrow (5.7, 7.1)$

95% CI on μ_{III}: $4.0000 \pm 2.048 \dfrac{\sqrt{0.9763}}{\sqrt{8}} \Rightarrow (3.3, 4.7)$

95% CI on μ_{IV}: $2.5000 \pm 2.048 \dfrac{\sqrt{0.9763}}{\sqrt{8}} \Rightarrow (1.8, 3.2)$

8.31
  a. The model for this experiment is given by
  $y_{ij} = \mu + \tau_i + \varepsilon_{ij}$, $i = 1, 2, 3$ and $j = 1, \ldots n_i$
  where $n_1 = 12, n_2 = 14, n_3 = 11$; μ = overall mean; τ_i = effect of ith division; ε_{ij} = random error associated with the jth response from the ith division
  b. Based on the ANOVA table below and a p-value of 0.001, we have sufficient evidence that there is a significant difference in average response among the divisions of the company.

```
Analysis of Variance

Source  DF  Adj SS  Adj MS  F-Value  P-Value
Factor   2   280.0  140.00     7.99    0.001
Error   34   595.7   17.52
Total   36   875.7
```

8.33

H_0: There is no difference in among the four varieties with respect to the yields.

H_a: At least one of the four varieties differs from the others with respect to the yields.

The Minitab output for the Kruskal-Wallis test is given here:

Kruskal-Wallis Test: Yields versus Variety

```
Kruskal-Wallis Test on Yields

Variety   N   Median   Ave Rank     Z
A         8    3.000      11.1   -1.89
B         8    3.750      21.9    1.87
C         8    4.200      24.4    2.74
D         8    2.850       8.7   -2.72
Overall  32               16.5

H = 16.50  DF = 3  P = 0.001
H = 16.56  DF = 3  P = 0.001  (adjusted for ties)
```

At the 0.05 significance level, we reject the null hypothesis since the p-value = 0.001 for the Kruskal-Wallis test. Thus, there is significant evidence that the distributions of yields differ for the four varieties. Both the F-test and H-test had p-values < 0.001, so the two tests produced equivalent results.

8.35

a. $H_0: \mu_A = \mu_B = \mu_C = \mu_D$ versus H_a: There is a difference in the means.

One-way ANOVA: Leaf Size vs Growing Condition
```
Source  DF      SS      MS      F       P
Factor   3   3961.2  1320.4   53.49   0.000
Error   36    888.7    24.7
Total   39   4849.8
```
$F = 53.49$ and p-value $< 0.001 \Rightarrow$ We reject the null hypothesis and conclude that there is significant evidence of a difference in the mean leaf size under the four growing conditions.

b. 95% CI on μ_A: $23.37 \pm 2.028 \dfrac{\sqrt{888.7/36}}{\sqrt{10}} \Rightarrow (20.18, 26.56)$

95% CI on μ_B: $8.58 \pm 2.028 \dfrac{\sqrt{888.7/36}}{\sqrt{10}} \Rightarrow (5.39, 11.77)$

95% CI on μ_C: $14.93 \pm 2.028 \dfrac{\sqrt{888.7/36}}{\sqrt{10}} \Rightarrow (11.74, 18.12)$

95% CI on μ_D: $35.15 \pm 2.028 \dfrac{\sqrt{888.9/36}}{\sqrt{10}} \Rightarrow (31.96, 38.34)$

The CI for the mean leaf size for Condition D implies that the mean is much larger for Condition D than for the other three conditions.

c. $H_0: \mu_A = \mu_B = \mu_C = \mu_D$ versus H_a: There is a difference in the means.

One-way ANOVA: Nicotine Content vs Growing Condition
```
Source  DF      SS      MS      F       P
Factor   3    18.08    6.03   2.10    0.117
Error   36   103.17    2.87
Total   39   121.25
```
$F = 2.10$ and p-value $= 0.117 \Rightarrow$ We fail to reject the null hypothesis and conclude that there is not significant evidence of a difference in the average nicotine content under the four growing conditions.

d. From the given data, it is not possible to conclude that the four growing conditions produce different average nicotine content.

e. No. If the testimony was supported by this experiment, then the test conducted in part (c) would have had the opposite conclusion.

8.37
a. Boxplots are given here:

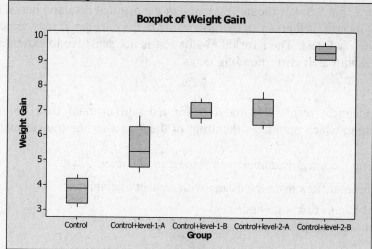

b. The summary statistics are given here:

Diet	n	Mean	Variance
Control	6	3.783	0.278
Control + Level 1 of A	6	5.500	0.752
Control + Level 2 of A	6	6.983	0.334
Control + Level 1 of B	6	7.000	0.128
Control + Level 2 of B	6	9.383	0.086

c. $F_{max} = \dfrac{0.752}{0.086} = 8.744 < 16.3$ (value from Table 12 with $\alpha = 0.05$) $\Rightarrow$ There is not significant evidence of a difference in the five variances. The boxplots do not reveal any deviations from the normality condition.

d. $H_0: \mu_1 = \mu_2 = \mu_3 = \mu_4 = \mu_5$ versus H_a: There is a difference in the means.

One-way ANOVA: Weight Gain versus Group
```
Source  DF       SS       MS      F      P
Group    4  103.038   25.760  81.67  0.000
Error   25    7.885    0.315
Total   29  110.923
```
$F = 81.67$ and p-value $< 0.001 \Rightarrow$ We reject the null hypothesis and conclude that there is significant evidence of a difference in the average weight gain under the five diets.

8.39

$H_0: \mu_A = \mu_B = \mu_C$ versus H_a: There is a difference in the means.

One-way ANOVA: Heights versus Herbicide
```
Source      DF       SS      MS      F      P
Herbicide    2   1146.3   573.2  39.14  0.000
Error       15    219.7    14.6
Total       17   1366.0
```
$F = 39.14$ and p-value $< 0.001 \Rightarrow$ We reject the null hypothesis and conclude that there is significant evidence of a difference in the average seedling height for the three herbicide groups.

76 Chapter 8: Inferences about More Than Two Population Central Values

8.41
The value of the Kruskal-Wallis statistic is identical to the value calculated prior to replacing 9.8 with 15.8. This will not happen in general, but 9.8 was the largest value in the original data and hence its rank would not be altered by increasing its size. If there is an extreme value in the data set, it may greatly alter the conclusion reached by the ANOVA F-test. The Kruskal-Wallis test is not sensitive to extreme values since it just replaces those extremes with their corresponding ranks.

8.43
The Kruskal-Wallis test yields identical results for the transformed and original data because the transformation was strictly increasing, which maintains the order of the data after the transformation has been performed.

H_0: There is no difference in among the three machines with respect to diameter.

H_a: At least one of the three machines differs from the others with respect to diameter.

The Minitab output for the Kruskal-Wallis test is given here:

Kruskal-Wallis Test: Diameter versus Machine

```
Kruskal-Wallis Test on Diameter

Machine    N   Median   Ave Rank      Z
1          5    7.500        5.4  -2.23
2          5    8.300        7.4  -1.35
3         10   20.650       14.6   3.10
Overall   20                10.5

H = 9.89   DF = 2   P = 0.007
```

At the 0.05 significance level, we reject the null hypothesis since the p-value = 0.007 for the Kruskal-Wallis test. Thus, there is significant evidence that the distributions of the diameters differ under the three machines.

The p-values from the F-test on the transformed data and from the Kruskal-Wallis test are nearly identical. Thus, we have equivalent conclusions about the differences in the diameters from the three machines.

Chapter 9

Multiple Comparisons

9.1
  a. The populations of interest are interviewers and job applicants with handicaps.
  b. The participating subjects were students. Thus, we cannot generalize results to those interviewers who would actually hire job applicants.

9.3
  a. $\hat{l}_1, \hat{l}_2,$ and $\hat{l}_4$ are contrasts
  b. $\hat{l}_2$ and $\hat{l}_4$ are orthogonal
  c. Testing these if these 2 contrasts are equal to 0 would be equivalent since $l_1 = 3l_2$

9.5
  a. $l_1 = 4\mu_C - \mu_H - \mu_A - \mu_B - \mu_W$
  b. $l_2 = 3\mu_H - \mu_A - \mu_B - \mu_W$
  c. $l_3 = \mu_A - 2\mu_B + \mu_W$
  d. $l_4 = \mu_B - \mu_W$

9.7 For purposes of this problem, the Midwest was considered to be 'in the West'
  a. $\hat{l}_1 = \mu_{NE} + \mu_{SE} - \mu_{MW} - \mu_W$
  b. $\hat{l}_2 = \mu_{NE} - \frac{1}{3}\mu_{SE} - \frac{1}{3}\mu_{MW} - \frac{1}{3}\mu_W$
  c. $\hat{l}_3 = \mu_{NE} - \mu_{SE}$
  d. Summary information below. All tests performed at the $\alpha = \frac{0.05}{3} = 0.016667$ level and a critical value of $F_{0.0166667,1,20} = 6.826$

Test	SSC	SSE	F	Decision
East vs. West	0.540563	0.001472	367.2986	Reject
Northeast vs. Others	1.711241	0.001472	1162.745	Reject
Northeast vs. Southeast	2.485125	0.001472	1688.58	Reject

  All three contrasts are highly significant.
  e. $\hat{l}_1$ and $\hat{l}_3$ are orthogonal but the other pairs are not. Therefore, the contrasts are not mutually orthogonal.

9.9 $\alpha_E \leq m\alpha_I \rightarrow \alpha_E \leq 8(0.005) \rightarrow \alpha_E = 0.04$

9.11
  a. $l = \frac{\mu_1 + \mu_2 + \mu_3}{3} - \frac{\mu_4 + \mu_5}{2}$
  Testing: $H_0: l \geq 0$ vs. $H_A: l < 0$
  $$SSC = 11644; MSE = 1112; F = 10.4 > F_{1,78,0.05} = 3.96$$
  Above results from software shows the families with 3 or fewer members have lower mean expenditures.

b. Because 10 tests were run, each was run at the individual alpha level of 0.005.

Fisher Pairwise Comparisons

```
Grouping Information Using the Fisher LSD Method and 99.5% Confidence

NumberMe   N    Mean    Grouping
5         10   131.90   A
4         14   124.9    A B
3         16   113.31   A B
2         23    98.65   A B
1         20    93.75     B

Means that do not share a letter are significantly different.
```
Only 1 and 5 member families have significantly different mean expenditures.

9.13 The boxplot indicates the distribution of the residuals is only slightly right skewed. This is confirmed with an examination of the normal probability plot. The Hartley test yields $F_{max} = 2.35 < 7.11$ using an $\alpha = 0.05$ test. Thus, the conditions needed to run the ANOVA F-test appear to be satisfied. From the output, $F = 15.68$, with p-value $< 0.0001 < 0.05$. Thus, we reject H_0 and conclude there is significant evidence of a difference in the average weight loss obtained using the five different agents.

9.15
  a. Critical value for Bonferroni is $t_{0.005, 45} = 2.7$

  If pairwise difference exceeds $LSD = 2.7 * \sqrt{\frac{2s_W^2}{10}} = 1.19$, it is significant.

  Significantly different pairs: (4, 3), (4, S), (1, 3), (1, S), (2, S)
  b. Scheffe's LSD $= \sqrt{0.9824 * 2/10}\sqrt{(5-1)F_{0.05, 4, 45}} = 1.42$
  Significantly different pairs: (4, 3), (4, S), (1, 3), (1, S), (2, S)
  c. All three methods find the same pairs being different.

9.17
  a. $l_1 = \mu_{A_1} + \mu_{A_2} + \mu_{A_3} + \mu_{A_4} - 4\mu_S$
  b. $l_2 = \mu_{A_1} - \mu_{A_2} + \mu_{A_3} - \mu_{A_4}$
  c. $l_3 = \mu_{A_1} + \mu_{A_2} - \mu_{A_3} - \mu_{A_4}$
  d. $l_4 = \mu_{A_1} + \mu_{A_3} - 2\mu_S$

9.19
  a. I will use the Tukey method although you can argue that Dunnett's is appropriate with continuous grazing as the control.

```
Tukey Simultaneous Tests for Differences of Means

Difference of      Difference    SE of                              Adjusted
Levels             of Means      Difference     95% CI      T-Value  P-Value
3-week - Cts        -0.547        0.432      (-1.649, 0.555)  -1.27   0.431
2-week - Cts        -0.867        0.432      (-1.969, 0.235)  -2.01   0.139
2-week - 3-week     -0.320        0.432      (-1.422, 0.782)  -0.74   0.743

Individual confidence level = 98.00%
```

Based on the CI's above (and p-values), there is no difference in mean soil density between continuous grazing and the other regimens in which there was no grazing period.

b. The confidence intervals above give insight into the size of the mean soil density differences among the regimens.

9.21 Using Fisher's LSD, we obtain

a.

Comparison	LSD	$\|\bar{y}_i - \bar{y}_j\|$	Conclusion
Manufacturing vs Marketing	1.707	7.4	Sig. Evidence Means are Diff.
Manufacturing vs Research	1.809	2.9	Sig. Evidence Means are Diff.
Marketing vs Research	1.748	4.5	Sig. Evidence Means are Diff.

Summary of Comparisons: <u>Manufacturing</u>   <u>Marketing</u>   <u>Research</u>

b. The employee acceptance appears to be higher in the marketing division than in the other two divisions with the research division significantly less than manufacturing.

9.23

a. The p-value from the F-test is 0.0345, which is less than 0.05; hence, there is significant evidence that the mean fat content in the four treatment groups are different.

b. Using a one-sided Dunnett's procedure: $D = 2.09\sqrt{\dfrac{2(0.1189)}{20}} = 0.228$

Comparison	D	$\bar{y}_i - \bar{y}_A$	Conclusion
B vs A	0.228	0.283	Sig. Evid. B's Mean is Greater than Control
C vs A	0.228	0.194	Not Sig. Evid. C's Mean is Greater than Control
D vs A	0.228	0.287	Sig. Evid. D's Mean is Greater than Control

9.25

a. $\hat{l} = 4593.2$, $SSC = \dfrac{20(4593.2)^2}{69.192} = 6,098,244.4$, $MS_{Error} = 1197.96 \Rightarrow$

$F = \dfrac{6098244.4}{1197.96} = 5090.5 \Rightarrow$ p-value $< 0.0001 \Rightarrow$

There is significant evidence that the contrast in the dose means is different from 0. Because $\hat{l}$ is positive, we conclude there is significant evidence that there is an increasing trend in the mean number of colonies as the dose level increases.

b. Yes, both the normality and equal variance assumptions appear to be violated. A transformation of the data should be made using either the square root or log transformation. Then conduct the F-test on the contrast in the mean of the transformed data.

80 Chapter 9: Multiple Comparisons

9.27

a. Using a one-sided Dunnett's procedure: $D = 2.16\sqrt{\dfrac{2(49.6685)}{44}} = 3.246$

Comparison	D	$\bar{y}_P - \bar{y}_i$	Conclusion
100 mg vs Placebo	3.246	0.7	Not Sig. evidence that 100 mg mean is greater than placebo
500 mg vs Placebo	3.246	3.9	Sig. evidence that 500 mg mean greater than placebo
1000 mg vs Placebo	3.246	2.1	Not Sig. evidence 1000 mg mean is greater than placebo
2000 mg vs Placebo	3.246	3.9	Sig. evidence that 2000 mg mean greater than placebo

b. Because a patient's blood pressure may change with just the knowledge that the patient is taking a pill that could affect blood pressure.

c. The coefficients in the contrast are $-800, -400, 100, 1100 \Rightarrow \hat{l} = -7000 \Rightarrow$

$$SSC = \dfrac{(-7000)^2}{\dfrac{(-800)^2}{44} + \dfrac{(-400)^2}{45} + \dfrac{(100)^2}{43} + \dfrac{(1100)^2}{44}} = 1069.09$$

$F = \dfrac{1069.09}{136.7825} = 7.82$ with df = 1, 216 $\Rightarrow$ 0.005 < p-value < 0.01 $\Rightarrow$

There is significant evidence of an increasing linear trend in the mean size of the reduction of change in blood pressure.

d. Using the SNK procedure:

$W_2 = 2.77\sqrt{\dfrac{136.78}{44}} = 4.88$, $W_3 = 3.31\sqrt{\dfrac{136.78}{44}} = 5.84$, $W_4 = 3.63\sqrt{\dfrac{136.78}{44}} = 6.40$

Comparison	W_r	$|\bar{y}_i - \bar{y}_j|$	Conclusion
100 mg vs 500 mg	4.88	5.9	Sig. Evidence Means are Diff.
100 mg vs 1000 mg	5.84	7.8	Sig. Evidence Means are Diff.
100 mg vs 2000 mg	6.40	7.8	Sig. Evidence Means are Diff.
500 mg vs 1000 mg	4.88	1.9	Not Sig. Evidence Means are Diff.
500 mg vs 2000 mg	5.84	1.9	Not Sig. Evidence Means are Diff.
1000 mg vs 2000 mg	4.88	0	Not Sig. Evidence Means are Diff.

Summary of Comparisons: <u>100 mg   500 mg   1000 mg   2000mg</u>

e. Neither the patients nor the medical personnel were aware of which dose levels were being given to the patients.

9.29

a. Based on the output below, the Kruskal-Wallis test suggests at least one pair of means differ.

Kruskal-Wallis Test: Dev versus Supplier

```
Kruskal-Wallis Test on Dev

Supplier    N   Median   Ave Rank      Z
1           9    189.1      32.8     2.50
2           9    174.1      23.0     0.00
```

© 2016 Cengage Learning. All Rights Reserved. May not be scanned, copied or duplicated, or posted to a publicly accessible website, in whole or in part.

```
3          9    156.9    14.0   -2.30
4          9    132.9     5.0   -4.60
5          9    204.4    40.2    4.40
Overall   45             23.0

H = 41.59   DF = 4   P = 0.000
H = 41.60   DF = 4   P = 0.000   (adjusted for ties)
```

$$KW_{ij} = \frac{3.86}{\sqrt{2}}\sqrt{\frac{7(7+1)}{12}\left(\frac{1}{7}+\frac{1}{7}\right)} = 3.1517$$

Comparison	$|R_i - R_j|$
1 vs 2	9.8
1 vs 3	18.8
1 vs 4	27.8
1 vs 5	7.4
2 vs 3	9
2 vs 4	18
2 vs 5	17.2
3 vs 4	9
3 vs 5	26.2
4 vs 5	35.2

All differences are statistically significant.

b. Pairwise difference significant if larger than $W^* = \frac{3.86}{\sqrt{2}}\sqrt{\frac{2*26.34}{7}} = 7.49$

All differences are statistically significant.

c. The nonparametric procedure is preferred because the conditions of ANOVA (equal variance and normally distributed data) were not fully satisfied in this data set.

9.31

a. Based on the output below, the Kruskal-Wallis test suggests at least one pair of medians differ.

Kruskal-Wallis Test: Discoloration versus Group

```
Kruskal-Wallis Test on Discoloration

Group     N   Median   Ave Rank      Z
I         8    8.000      27.7     3.89
II        8    6.250      21.1     1.61
III       8    4.000      11.8    -1.65
IV        8    2.250       5.4    -3.85
Overall  32               16.5

H = 26.50  DF = 3  P = 0.000
H = 26.62  DF = 3  P = 0.000  (adjusted for ties)
```

$$KW_{ij} = \frac{3.63}{\sqrt{2}}\sqrt{\frac{8(8+1)}{12}\left(\frac{1}{8}+\frac{1}{8}\right)} = 3.1437$$

Comparison	$|R_i - R_j|$
I vs II	6.6
I vs III	15.9
I vs IV	22.3
II vs III	9.3
II vs IV	15.7
III vs IV	6.4

All differences are statistically significant.

b. Because there were not any severe violations of our assumptions, the Tukey procedure would be preferred due to higher power.

Chapter 10

Categorical Data

10.1

a. $\hat{\pi} \pm z_{\alpha/2}\sqrt{\frac{\hat{\pi}(1-\hat{\pi})}{n}} = 0.35 \pm 2.58\sqrt{\frac{0.35(0.65)}{20}} = (0.0753, 0.62517)$

WAC CI: $\tilde{y} = y + 0.5z_{\alpha/2}^2 = 10.3282$

$\tilde{n} = n + z_{\alpha/2}^2 = 25.6564, \tilde{\pi} = \frac{\tilde{y}}{\tilde{n}} = 0.4026$

$$\tilde{\pi} \pm z_{\alpha/2}\sqrt{\frac{\tilde{\pi}(1-\tilde{\pi})}{\tilde{n}}} = (0.1440, 0.6309)$$

The WAC procedure is preferred over the standard CI, except for exceptionally large sample sizes.

b. Following the same procedure as above. WAC adjustment preferred.

	Lower Limit	Upper Limit
Large Sample Approx	0.6256	0.9744
WAC	0.5794	0.9247

c. Following the same procedure as above. WAC adjustment likely not needed.

	Lower Limit	Upper Limit
Large Sample Approx	0.1671	0.5138
WAC	0.1944	0.523

d. Following the same procedure as above. WAC adjustment likely not needed.

	Lower Limit	Upper Limit
Large Sample Approx	0.0362	0.2038
WAC	0.0561	0.2313

10.3

a. Following the same procedure as above. WAC adjustment preferred.

	Lower Limit	Upper Limit
Large Sample Approx	0.2171	0.7829
WAC	0.255	0.745

b. Following the same procedure as above. WAC adjustment preferred.

	Lower Limit	Upper Limit
Large Sample Approx	0.0432	0.3568
WAC	0.0841	0.3958

c. Following the same procedure as above. WAC adjustment needed.

	Lower Limit	Upper Limit
Large Sample Approx	0.023	0.227
WAC	0.0488	0.269

Chapter 10: Categorical Data

d. Following the same procedure as above. WAC adjustment needed.

	Lower Limit	Upper Limit
Large Sample Approx	0.0073	0.0927
WAC	0.0182	0.1164

10.5

a. $n = \frac{z_{\alpha/2}^2 \pi(1-\pi)}{E^2} = \frac{1.96^2(0.5)(0.5)}{0.04^2} = 600.25 \rightarrow 601$

b. $n = \frac{z_{\alpha/2}^2 \pi(1-\pi)}{E^2} = \frac{1.96^2(0.3)(0.7)}{0.04^2} = 504.21 \rightarrow 505$

c. In part (a), we have no prior knowledge for an estimate of the population proportion. Thus, the sample size we get is the maximum sample size needed to guarantee a margin of error of 0.04. In part (b), we have *a priori* information about the population proportion. Since 0.30 is closer to 0 than 0.50, we get a smaller sample size. In other words, we need to sample fewer people when the population proportion is closer to 0 (or to 1) than when the population proportion is unknown.

10.7

a. $\hat{\pi} = \frac{591}{10,000} = 0.0591;$

$$\hat{\pi} \pm z_{\alpha/2}\sqrt{\frac{\hat{\pi}(1-\hat{\pi})}{n}} = 0.0591 \pm 2.58\sqrt{\frac{0.0591(0.9409)}{10,000}} = (0.05304, 0.06516)$$

We are 99% confident that the interval 0.05304 to 0.06516 contains the true proportion of false positives produced by the test.

b. Since 0.05 (5%) is entirely below the 99% confidence interval, we do not have evidence that the test produces less than 5% false positives at the $\alpha = 0.01$ level. (We have evidence that the test produces more than 5% false positives.)

10.9

a. $P(AIDS|positive\ test) = \frac{P(AIDS\ and\ positive\ test)}{P(positive\ test)} = \frac{993}{1584} = 0.6269$

b. $P(no\ AIDS|negative\ test) = \frac{P(AIDS\ and\ negative\ test)}{P(negative\ test)} = \frac{9409}{9416} = 0.9993$

10.11 $n = \frac{z_{\alpha/2}^2 \pi(1-\pi)}{E^2} = \frac{2.58^2(0.5)(0.5)}{0.05^2} = 665.64 \rightarrow 666$

10.13

a. Test $H_0: \pi \geq 0.1\ vs.\ H_A: \pi < 0.1$

Because $n\pi = 4 < 5$, we need to do a test based on the binomial distribution (no normal approximation).

$$p - value = 1 - P(Y \leq 5) = 0.794$$

Therefore, we do not have sufficient evidence that less than 10% are dissatisfied. This is clear because the data has more than 10% dissatisfied customers.

b. $\tilde{y} = y + 0.5z_{\alpha/2}^2 = 5 + 0.5(1.96)^2 = 6.92$

$\tilde{n} = n + z_{\alpha/2}^2 = 41.92, \tilde{\pi} = \frac{\tilde{y}}{\tilde{n}} = 0.165$

$\tilde{\pi} \pm z_{\alpha/2}\sqrt{\frac{\tilde{\pi}(1-\tilde{\pi})}{\tilde{n}}} = 0.165 \pm 1.96\sqrt{\frac{0.165(0.835)}{41.92}} = (0.053, 0.277)$

10.15

a. $\hat{\pi} = \frac{230}{1500} = 0.1533$; $\hat{\pi} \pm z_{\alpha/2}\sqrt{\frac{\hat{\pi}(1-\hat{\pi})}{n}} = 0.1533 \pm 1.96\sqrt{\frac{0.1533(0.8467)}{1500}} = (0.1351, 0.1715)$ We are 95% confident that the interval 0.1351 to 0.1715 contains the true proportion of registered voters who would favor drilling for oil in national parks.

b. $n = \frac{z_{\alpha/2}^2 \pi(1-\pi)}{E^2} = \frac{(1.96)^2(0.1533)(0.8467)}{(0.01)^2} = 4986.4 \Rightarrow n = 4987$

c. Let π be the proportion of registered voters who would favor drilling for oil in national parks.

$H_0: \pi \leq 0.50$ versus $H_a: \pi > 0.50$

For $\alpha = 0.05$, reject H_0 if $z > 1.645$.

$n\pi_0 = n(1-\pi_0) = 1500(0.5) = 750 > 5$

$z = \frac{\hat{\pi} - \pi_0}{\sigma_{\hat{\pi}}} = \frac{0.1533 - 0.50}{\sqrt{\frac{0.50(0.50)}{1500}}} = -26.86 < 1.645$ with p-value $\approx 1.0 > 0.05$.

Thus, we fail to reject H_0. There is not sufficient evidence to conclude that more than half of registered voters who would favor drilling for oil in national parks.

10.17 $z_{\alpha/2}\sigma_{\hat{\pi}_A - \hat{\pi}_B} = 2.33\sqrt{\frac{0.3(0.7)}{n} + \frac{0.15(0.85)}{n}} = 0.01 \Rightarrow 2.33^2\left(\frac{0.3(0.7)}{n} + \frac{0.15(0.85)}{n}\right) = 0.01^2.$

$\Rightarrow n = \frac{2.33^2(0.3(0.7) + 0.15(0.85))}{0.01^2} = 18322.5375 \rightarrow n = 18323$

We need to sample 18,323 observations from each population.

10.18

a. From Minitab (using the normal approx.):

```
95% CI for difference:  (0.000146530, 0.479853)
```

We are 95% confident that the proportion of customers who will buy the lawn mower increases by between 0.0001465 and 0.48 when offered the warranty.

b. Denote the proportion of all customers who will buy a mower who are offered a warranty by π_1 and who are not offered a warranty by π_2.

$H_0: \pi_1 - \pi_2 \leq 0$ versus $H_a: \pi_1 - \pi_2 > 0$

Test and CI for Two Proportions

```
Sample   X    N    Sample p
1        10   25   0.400000
2         4   25   0.160000
Difference = p (1) - p (2)
Estimate for difference:  0.24
95% lower bound for difference:  0.0387086
Test for difference = 0 (vs > 0):  Z = 1.96  P-Value = 0.025
```

We reject H_0 (since $p-value < 0.05$) and conclude that there is significant evidence that offering the warranty will increase the proportion of customers who will purchase a mower.

c. The conditions are not met $(n_2\hat{\pi}_2 = 4)$, so a Fisher's exact test is more appropriate.

```
Fisher's exact test: P-Value = 0.057
```
Fisher's exact test fails to reject (p>0.05) so the warranty offer does not significantly increase probability of purchase.

d. Based on the results above, the dealer should not offer the warranty with the information provided, however, it is recommended that more data be collected to potentially show the impact of the warranty.

10.19
a. From Minitab (using the normal approx.):

```
95% CI for difference:  (-0.277253, -0.00274653)
```

We are 95% confident that between 0.00275% and 27.73% more females prefer the color advertisement.

b. Denote the proportion of all customers who will buy a mower who are offered a warranty by π_1 and who are not offered a warranty by π_2.

$H_0: \pi_M - \pi_F = 0$ versus $H_0: \pi_M - \pi_F \neq 0$

Test and CI for Two Proportions

```
Sample   X    N    Sample p
1        39   50   0.780000
2        46   50   0.920000


Difference = p (1) - p (2)
Estimate for difference:  -0.14
95% CI for difference:  (-0.277253, -0.00274653)
Test for difference = 0 (vs ≠ 0):  Z = -2.00  P-Value = 0.046
```

We reject H_0 (since $p-value < 0.05$) and conclude that there is significant evidence the proportion of males who prefer the color advertisement is different from the proportion of females who prefer the color advertisement.

c. The conditions are not met ($n_2(1 - \hat{\pi}_F) = 4$), so a Fisher's exact test is more appropriate.

```
Fisher's exact test: P-Value = 0.09
```
Fisher's exact test fails to reject (p>0.05) so there is not a significant difference in preference for the color advertisement between males and females.

d. Based on the results above, the firm should not use different ads for male and female customers, but more data would be warranted to be completely sure.

10.21

a. New: $\tilde{y} = y + 0.5z_{\alpha/2}^2 = 5 + 0.5(1.96)^2 = 6.9208$;

$\tilde{n} = n + z_{\alpha/2}^2 = 50 + (1.96)^2 = 53.8416$; $\tilde{\pi} = \dfrac{\tilde{y}}{\tilde{n}} = \dfrac{6.9208}{53.8416} = 0.1285$

$\tilde{\pi} \pm z_{\alpha/2}\sqrt{\dfrac{\tilde{\pi}(1-\tilde{\pi})}{\tilde{n}}} = 0.1285 \pm 1.96\sqrt{\dfrac{0.1285(0.8715)}{53.8416}} = (0.0391, 0.2179)$

Old: $\tilde{y} = y + 0.5z_{\alpha/2}^2 = 9 + 0.5(1.96)^2 = 10.9208$;

$\tilde{n} = n + z_{\alpha/2}^2 = 50 + (1.96)^2 = 53.8416$; $\tilde{\pi} = \dfrac{\tilde{y}}{\tilde{n}} = \dfrac{10.9208}{53.8416} = 0.2028$

$\tilde{\pi} \pm z_{\alpha/2}\sqrt{\dfrac{\tilde{\pi}(1-\tilde{\pi})}{\tilde{n}}} = 0.2028 \pm 1.96\sqrt{\dfrac{0.2028(0.7972)}{53.8416}} = (0.0954, 0.3102)$

b. Denote the proportion of all plants with the new treatment that have a toxic level of nickel by π_1 and the proportion of all plants with the old treatment that have a toxic level of nickel by π_2.

$H_0: \pi_1 - \pi_2 \geq 0$ versus $H_a: \pi_1 - \pi_2 < 0$

$z = \dfrac{0.10 - 0.18}{\sqrt{\dfrac{0.10(0.90)}{50} + \dfrac{0.18(0.82)}{50}}} = -1.16 > -1.645$ with p-value = 0.1230 > 0.05.

We fail to reject H_0 and conclude that there is not significant evidence that the new treatment would have a lower proportion of plants having a toxic level of nickel.

c. p-value = $P(X \leq 5)$
$= P(X=0) + P(X=1) + P(X=2) + P(X=3) + P(X=4) + P(X=5)$

$= \dfrac{\binom{50}{0}\binom{50}{14}}{\binom{100}{14}} + \dfrac{\binom{50}{1}\binom{50}{13}}{\binom{100}{14}} + \dfrac{\binom{50}{2}\binom{50}{12}}{\binom{100}{14}} + \dfrac{\binom{50}{3}\binom{50}{11}}{\binom{100}{14}} + \dfrac{\binom{50}{4}\binom{50}{10}}{\binom{100}{14}} + \dfrac{\binom{50}{5}\binom{50}{9}}{\binom{100}{14}}$

$= 0.00002 + 0.0004 + 0.0034 + 0.0166 + 0.0535 + 0.1201$

$= 0.1940$

Like in part (b), we fail to reject H_0 and conclude that the new treatment would not have a lower proportion of plants having a toxic level of nickel.

d. The 95% CI is $(0.10-0.18) \pm 1.96\sqrt{\frac{0.10(0.90)}{50}+\frac{0.18(0.82)}{50}} = (-0.215, 0.055)$. Note that 0 is contained in the CI, so we would again fail to reject H_0.

10.23
    a. The proportion of correct identifications by the two systems are correlated because they are performed on a benchmark of 2000 words. The table shows how each word is identified in both of the systems.

    b. $H_0: \pi_{1.} = \pi_{.1}$ versus $H_A: \pi_{1.} \neq \pi_{.1}$
Since $n_{12} + n_{21} > 20$, we can use the large sample test.
$$z = \frac{n_{12} - n_{21}}{\sqrt{n_{12} + n_{21}}} = \frac{58 - 16}{\sqrt{58 + 16}} = 4.88 > 1.96$$
We have sufficient evidence that there is a difference in the performances of the two systems.

    c. $\hat{\pi}_{1.} - \hat{\pi}_{.1} \pm z_{\alpha/2} \frac{1}{m}\sqrt{(n_{12} + n_{21}) - \frac{1}{m}(n_{12} - n_{21})^2}$

$$0.9895 - 0.9685 \pm 1.96 \frac{1}{2000}\sqrt{74 - \frac{1}{2000}(42)^2} = (0.0126, 0.0294)$$

10.25 A multinomial experiment has the following characteristics:
1. The experiment consists of n identical trials.
2. Each trial results in one of k outcomes.
3. The probability that a single trial will result in outcome i is π_i, $i = 1, \ldots, k$, and remains constant from trial to trial. (Note: $\sum_i \pi_i = 1$)
4. The outcomes of the trials are independent.
5. We observe n_i, the number of the n trials resulting in outcome i. (Note: $\sum_i n_i = n$)

10.27 The goodness-of-fit test for a multinomial experiment has approximately a chi-square distribution. The approximation is adequate if no expected cell count E_i is less than 1 and no more than 20% of the E_is are less than 5. When H_0 is not rejected, the probability of a Type II error may have been committed. The calculation of this error is very difficult. Thus, the researcher must be very cautious concerning any test result in which the null hypothesis has not been rejected. When we reject H_0, we know with certainty that the probability of a Type I error is at most α, but we do not have this level of certainty concerning the chance of a Type II error when we fail to reject H_0, especially when $n\pi_i$ is small or k is large.

10.29

$$H_0: \pi_1 = 0.8; \pi_2 = 0.1; \pi_3 = 0.05; \pi_4 = 0.03; \pi_5 = 0.02$$
$$H_A: \text{At least one } \pi_i \text{ differs from above}$$

	Observed	Expected	Contribution=$\frac{(n-E)^2}{E}$
0	238	240	0.016667
1	32	30	0.133333
2	12	15	0.6
3	13	9	1.7778
4 or more	5	6	0.16667
Total	300	300	2.694

$$p - \text{value} = P(\chi^2_4 \geq 2.694) = 0.61$$

Therefore, we do not have sufficient evidence to conclude a significant change in the proportion of defectives.

10.31
  a. The population of interest is all horses that race in these (or similar races) with 8 starting gates.
  b. $H_0: \pi_i = \frac{1}{8}$ for each gate versus $H_A:$ At least one $\pi_i \neq \frac{1}{8}$

Chi-Square Goodness-of-Fit Test for Observed Counts in Variable: Num Winners

```
                    Test                 Contribution
Category  Observed  Proportion  Expected  to Chi-Sq
1           29        0.125       18      6.72222
2           19        0.125       18      0.05556
3           18        0.125       18      0.00000
4           25        0.125       18      2.72222
5           17        0.125       18      0.05556
6           10        0.125       18      3.55556
7           15        0.125       18      0.50000
8           11        0.125       18      2.72222


  N   DF   Chi-Sq   P-Value
144    7  16.3333    0.022
```

Based on the p-value of 0.022, there is sufficient evidence that the starting position has some effect on the chance of winning the race.

10.33
  a. The following bar chart shows the juror age groups as a percent of the total:

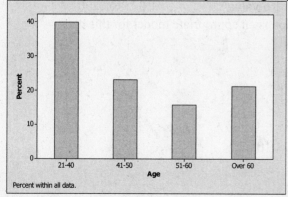

b. $H_0: \pi_1 = 0.421, \pi_2 = 0.229, \pi_3 = 0.157, \pi_4 = 0.193$

H_a: at least one of the π_is differs from its hypothesized value

$E_i = n\pi_{i0} \Rightarrow$

$E_1 = 1000(0.421) = 421$, $E_2 = 1000(0.229) = 229$, $E_3 = 1000(0.157) = 157$,

$E_4 = 1000(0.193) = 193$

$\chi^2 = \sum_{i=1}^{4} \frac{(n_i - E_i)^2}{E_i} = 3.044$ with df = 4 – 1 = 3 $\Rightarrow$ p-value = 0.3849 $\Rightarrow$ We fail to reject H_0. It does not appear that there is a difference between the age distribution of jurors and the countywide age distribution.

c. Since we failed to reject the null hypothesis, there does not appear to be any age bias in the selection of jurors.

10.35

$H_0: \pi_1 = 0.0625, \pi_2 = 0.25, \pi_3 = 0.375, \pi_4 = 0.25, \pi_5 = 0.0625$

H_a: at least one of the π_is differs from its hypothesized value

$E_i = n\pi_{i0} \Rightarrow E_1 = 125(0.0625) = 7.8125$, $E_2 = 125(0.25) = 31.25$,

$E_3 = 125(0.375) = 46.875$, $E_4 = 125(0.25) = 31.25$, $E_5 = 125(0.0625) = 7.8125$

$\chi^2 = \sum_{i=1}^{5} \frac{(n_i - E_i)^2}{E_i} = 7.608$ with df = 5 – 1 = 4 $\Rightarrow$ p-value = 0.1070 $\Rightarrow$ We fail to reject H_0. The data appear to fit the hypothesized theory that the securities analysts perform no better than chance. However, we have no indication of the probability of a Type II error.

10.37

a. Based on this data, $\hat{\lambda} = 0.8$

Breaks/Bar	Freq	Expected	Contribution
0	121	125.8121	0.18405543
1	110	100.6497	0.8686399
2	38	40.25988	0.12685176
3	7	10.73597	1.30006432
4	3	2.147193	0.33871155
5	1	0.343551	1.25432747
	280	X^2	4.07265044
		p-value	0.39626296

These data suggest the Poisson distribution is an appropriate model for this data.

b. All nylon bars produced via this process.

10.39

a. From the data, $\bar{y} \approx \frac{1}{100} \sum_i (n_i)(y_i) = 5.57$

$s^2 \approx \frac{1}{99} \sum_i (n_i)(y_i - 5.57)^2 = 1056.5/99 = 10.67$

b. Using $\mu = 5.5$, the Poisson table yields the following probabilities after combining the first two categories and combining the last four categories so that $E_i > 1$ and only one E_i is less than 5:

k	≤ 1	2	3	4	5	6	7	8	≥ 9
$\pi_i = P(y=k)$	0.0266	0.0618	0.1133	0.1558	0.1714	0.1571	0.1234	0.0849	0.1057
$E_i = 100\pi_i$	2.66	6.18	11.33	15.58	17.14	15.71	12.34	8.49	10.57

$$\chi^2 = \sum_{i=1}^{9} \frac{(n_i - E_i)^2}{E_i} = 13.441 \text{ with df} = 9 - 2 = 7 \Rightarrow \text{p-value} = 0.0621 \Rightarrow \text{We fail to reject } H_0.$$

The conclusion that the number of fire ant hills follows a Poisson distribution appears to be supported by the data. However, we have not computed the probability of making a Type II error, so the conclusion is somewhat tenuous.

c. The fire ant hills are somewhat more clustered than randomly distributed across the pastures, although the data failed to reject the null hypothesis that the fire ant hills were randomly distributed.

10.41
a.

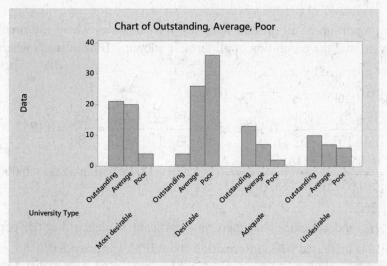

b. It appears the employees coming from the 'most desirable' schools yielded the most outstanding performance reviews, but the employees coming from 'adequate' and 'poor' schools had a higher percentage of outstanding reviews. Perhaps those from lower level schools showed more work-ethic. For some reason, 'desirable' schools had very low performance rating employees relatively.

10.43
a.

Chi-Square Test for Association: Promotion Decision, Worksheet columns
```
Rows: Promotion Decision   Columns: Worksheet columns

                    39 or    40 or
                    Younger  Older   All

Promoted               38      46     84
                    39.97   44.03
```

```
Not Promoted           80      84    164
                    78.03   85.97

All                   118     130    248

Cell Contents:        Count
                      Expected count
```

```
Pearson Chi-Square = 0.279, DF = 1, P-Value = 0.597
Likelihood Ratio Chi-Square = 0.280, DF = 1, P-Value = 0.597
```

Therefore, there is not sufficient evidence that the promotion decision is related to the age of the employee at when considered at two levels.

b. The results differ. This is likely due to the few promotions for the very young (30 and younger) or very old (50 and older).

10.45

a. H_0: The percentage of farmers adopting IPM is not increased when information about IPM was provided.

H_a: The percentage of farmers adopting IPM is increased when information about IPM was provided.

Denote the proportion of all farmers adopting IPM in states where information about IPM was provided by π_1 and the proportion of all farmers adopting IPM in states where information about IPM was not provided by π_2.

$H_0: \pi_1 - \pi_2 \leq 0$ versus $H_a: \pi_1 - \pi_2 > 0$

$$\hat{\pi}_1 = \frac{39+55+30}{131+124+107} = \frac{124}{362} = 0.3425 \; ; \; \hat{\pi}_2 = \frac{19+22+24}{133+110+107} = \frac{65}{350} = 0.1857$$

$$z = \frac{0.3425 - 0.1857}{\sqrt{\frac{0.3425(0.6575)}{362} + \frac{0.1857(0.8143)}{350}}} = 4.829 > 1.645 \text{ with p-value} < 0.0001.$$

We reject H_0 and conclude that there is significant evidence that the percentage of farmers adopting IPM is increased when information about IPM was provided.

b. It appears that providing information about IPM is associated with higher adoption rates. Thus, if the program wants to expand, agents should discuss IPM with grape farmers in other states. (Note this is an association and causation should not be inferred.)

10.47

H_0: Graduation rate is independent of stepparent family type.

H_a: Graduation rate depends on stepparent family type.

$$\chi^2 = \sum_{i,j} \frac{(n_{ij} - E_{ij})^2}{E_{ij}} = 10.332 \text{ with df} = (2-1)(2-1) = 1, \text{ p-value} = 0.0013. \text{ We reject } H_0. \text{ There is}$$

significant evidence that the graduation rates depend on stepparent family type.

10.49

It appears that those between 30 and 49 tend to be promoted at a higher rate than those under 30 and those 50 and over; whereas the middle manager who were not promoted are equally distributed over the four age groups.

10.51
P(Graduated = Yes given Family = Two-Parent) = 407/452 = 0.9004
P(Graduated = Yes given Family = Single-Always) = 61/77 = 0.7922
P(Graduated = Yes given Family = Single-Divorced) = 231/260 = 0.8885
P(Graduated = Yes given Family = No Cohab) = 124/135 = 0.9185
P(Graduated = Yes given Family = With Cohab) = 193/244 = 0.7910

P(Graduated = No given Family = Two-Parent) = 45/452 = 0.0996
P(Graduated = No given Family = Single-Always) = 16/77 = 0.2078
P(Graduated = No given Family = Single-Divorced) = 29/260 = 0.1115
P(Graduated = No given Family = No Cohab) = 11/135 = 0.0815
P(Graduated = No given Family = With Cohab) = 51/244 = 0.2090

The most notable feature is that it appears that stepparent families where the parents live together and always single-parent families have a high rate of students not graduating from high school.

10.53

a. $\hat{\mu}_{\ln(OR)} = \ln\left(\dfrac{0.636/0.364}{0.312/0.688}\right) = 1.349$ and $\hat{\sigma}_{\ln(OR)} = \sqrt{\dfrac{1}{159} + \dfrac{1}{91} + \dfrac{1}{78} + \dfrac{1}{172}} = 0.1895$

$z = \dfrac{1.349 - 0}{0.1895} = 7.12 \Rightarrow$ p-value < 0.0001

We reject H_0. There is a significant difference between the low and high dietary cholesterol intake groups relative to their risk of having high blood pressure.

b. $\ln(OR) \pm z_{\alpha/2} \hat{\sigma}_{\ln(OR)} = 1.349 \pm 1.96(0.1895) = (0.97758, 1.72042)$ The 95% CI for the odds ratio is $(e^{0.97758}, e^{1.72042}) = (2.658, 5.587)$. Because the 95% confidence interval does not include an odds ratio of 1, we may conclude that there is a statistically detectable relation between dietary cholesterol and high blood pressure.

c. Yes, the conclusions are consistent.

10.55

a. Seat belt: $\dfrac{16,001/7,774,635}{1-(16,001/7,774,635)} = 0.00206$;

no seat belt: $\dfrac{31,199/2,823,086}{1-(31,199/2,823,086)} = 0.0112$.

The odds of being killed in a harmful event car accident while wearing a seat belt are about 444% lower than the odds of being killed in a harmful event car accident for a vehicle while not wearing a seat belt.

b. $OR = \dfrac{0.0021/0.9979}{0.0111/0.9889} = 0.188$ The odds of being killed in a harmful event car accident is 81.2% lower for those wearing seat belts for those not wearing seat belts. This means that wearing a seat belt may decrease your chance of dying in a harmful event car accident.

94 Chapter 10: Categorical Data

c. $\hat{\mu}_{\ln(OR)} = \ln\left(\dfrac{0.0021/0.9979}{0.0111/0.9889}\right) = -1.674$ and

$$\hat{\sigma}_{\ln(OR)} = \sqrt{\dfrac{1}{16,001} + \dfrac{1}{31,199} + \dfrac{1}{7,758,634} + \dfrac{1}{2,791,887}} = 0.0097$$

$z = \dfrac{-1.674 - 0}{0.0097} = -172.57 \Rightarrow$ p-value < 0.0001

We reject H_0. There is a significant difference between wearing a seat belt and not wearing a seat belt relative to their risk of a person being killed in a harmful event car accident.

d. $\ln(OR) \pm z_{\alpha/2}\hat{\sigma}_{\ln(OR)} = -1.674 \pm 1.96(0.0097) = (-1.693, -1.655)$ The 95% CI for the odds ratio is $(e^{-1.693}, e^{-1.655}) = (0.184, 0.191)$. Because the 95% confidence interval does not include an odds ratio of 1, we may conclude that there is a statistically detectable relation between wearing and not wearing a seat belt and being killed in a harmful event car accident.

10.57

a. H_0: Proper usage of seat belts is independent of air bag installation.

H_a: Proper usage of seat belts depends on air bag installation.

$$\chi^2 = \sum_{i,j} \dfrac{(n_{ij} - E_{ij})^2}{E_{ij}} = 817,875.961$$ with df = $(2-1)(2-1) = 1$, p-value < 0.0001. We reject H_0. There is significant evidence that proper usage of seat belts depends on air bag installation. Note that the chi-square statistic is ridiculously large because of the large number of people in the study.

b. A 95% CI for the difference in those who wear seat belts and those who do not wear seat belts who have air bags installed is

$$(0.6266 - 0.4461) \pm 1.96\sqrt{\dfrac{0.6266(0.3734)}{7,774,635} + \dfrac{0.4461(0.5539)}{2,823,086}} = (0.1798, 0.1812).$$

It appears that about 18% more people who have air bags use seat belts than those who do not wear seat belts.

10.59

a. There are 1200 total batteries in the 300 flashlights.

$$\hat{\pi}_i = \dfrac{100*0 + 126*1 + 60*2 + 13*3 + 1*4}{1200} = \dfrac{289}{1200} = 0.2408$$

b. $\hat{\pi} \pm z_{\alpha/2}\sqrt{\dfrac{\hat{\pi}(1-\hat{\pi})}{n}} = 0.2408 \pm 1.96\sqrt{\dfrac{0.2408(0.7592)}{1200}} = (0.217, 0265)$

c. Because the interval doesn't include 0.15 (it's entirely above it), there is sufficient evidence to refute the claim that at most 15% of its batteries are defective.

10.61

a. Let π_{i0} be the probability of winning when starting in position i.

H_0: $\pi_{10} = \pi_{20} = \cdots = \pi_{70} = 1/7$

H_a: Not all π_{i0}'s equal 1/7.

b. $E_i = n\pi_{i0} \Rightarrow E_1 = E_2 = E_3 = E_4 = E_5 = E_6 = E_7 = 144(1/7) = 20.57$,

$\chi^2 = \sum_{i=1}^{7} \frac{(n_i - E_i)^2}{E_i} = 16.863$ with df = 7 – 1 = 6 $\Rightarrow$ p-value = 0.0097 $\Rightarrow$ We reject H_0. There appears to be a difference in the chance of winning based on starting position.

10.63

a. $H_0: \pi_1 - \pi_2 = 0 \ vs. \ H_A: \pi_1 - \pi_2 \neq 0$

$$z = \frac{0.855 - 0.775 - 0}{\sqrt{\frac{0.855(0.145)}{200} + \frac{0.775(0.225)}{200}}} = 2.07$$

$p - value = 0.019 < 0.05$

Therefore, we have sufficient evidence that level of satisfaction is different between the methods.

b. $0.855 - 0.775 \pm 1.96\sqrt{\frac{0.855(0.145)}{200} + \frac{0.775(0.225)}{200}} = (0.0043, 0.1557)$

The interval is consistent with the hypothesis test as 0 is not in the confidence interval.

10.65

Test and CI for Two Proportions

```
Sample   X    N    Sample p
1        65   100  0.650000
2        59   100  0.590000


Difference = p (1) - p (2)
Estimate for difference:  0.06
95% lower bound for difference:  -0.0526935
Test for difference = 0 (vs > 0):  Z = 0.88  P-Value = 0.191

Fisher's exact test: P-Value = 0.233
```

a. Based on the minitab output, there is not significant the new pesticide is more effective than the old ($p - value = 0.191 > 0.05$).
b. Fisher's exact test yields a p-value of 0.233 which agrees with part a (the new pesticide isn't significantly more effective).
c. 95% CI for difference: $(-0.074, 0.194)$
d. None of the tests in a-c provide evidence that the new formulation is more effective.

10.67

a. Under the hypothesis of independence, the expected frequencies are given in the following table:

$\hat{E}_{ij} = \frac{n_{i\cdot}n_{\cdot j}}{900}$

	Opinion				
Commercial	1	2	3	4	5
A	42	107	78	34	39
B	42	107	78	34	39
C	42	107	78	34	39

b. df = (3 – 1)(5 – 1) = 8

c. The cell chi-squares are given in the following table:

Commercial	Opinion				
	1	2	3	4	5
A	2.3810	3.7383	2.1667	4.2353	0.6410
B	2.8810	10.8037	0.0513	5.7647	21.5641
C	0.0238	1.8318	1.5513	0.1176	14.7692

$$\chi^2 = \sum_{i,j} \frac{(n_{ij} - E_{ij})^2}{E_{ij}} = 72.521$$ with df = 8 ⇒ p-value < 0.0001 ⇒ We reject H_0. There is significant evidence that the commercial viewed and opinion are related.

10.69

a. $H_0: \pi_{1.} = \pi_{.1}$ versus $H_A: \pi_{1.} \neq \pi_{.1}$

Since $n_{12} + n_{21} < 20$, we should use the exact binomial test which yields a p-value of approx. 1. Since this is greater than 0.05, we do not have sufficient evidence that the drugs differ in effectiveness.

b. The Minitab output yields:
```
McNemar's Test

Estimated
Difference          95% CI              P
  0.0222      (-0.1897, 0.2342)      1.000
```

10.71

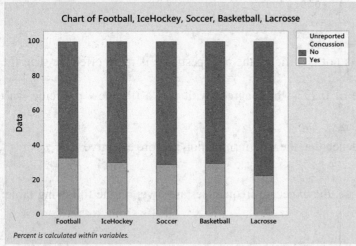

The graph shows the proportion of unreported concussions were approximately the same across sport.

10.73
Chi-Square Test for Association: Frequency, Worksheet columns

```
Rows: Frequency    Columns: Worksheet columns

            Poor   Average    Good  Excellent   All

1              3         4      37         44    88
            3.929    11.786  34.571     37.714

2              2         6      30         28    66
            2.946     8.839  25.929     28.286

3              3         8      16         19    46
            2.054     6.161  18.071     19.714

4 or more      2        12       5          5    24
            1.071     3.214   9.429     10.286

All           10        30      88         96   224

Cell Contents:        Count
                      Expected count


Pearson Chi-Square = 39.303, DF = 9, P-Value = 0.000
Likelihood Ratio Chi-Square = 32.021, DF = 9, P-Value = 0.000

* NOTE * 5 cells with expected counts less than 5
```

a. The p-value < 0.05 which shows significant evidence that the frequency of purchase is related to the adequacy of selection, but the conditions are suspect.

b. It appears the stores with "excellent" selection are more likely to yield a frequency of purchase of 1. "Average" adequacy yields the largest frequency of purchase which may be related to specialty stores.

10.75 The counts and adequacy rating percentages (row percentages) for each frequency level are given in the table contained in the solution for Exercise 10.74. It appears that when frequency use is lower, the adequacy ratings tend to be higher. For instance, when frequency use is 1, over 92% of respondents gave a rating of 3 or 4. Although there is slight drop off, low frequency users tend to give a high selection rating. This trend carries over into frequency use category 3, where 87.8% of respondents gave a 3 or 4 rating. Only in the highest frequency of use category does this trend reverse itself. One possible explanation for this result is perhaps these people have read so many books that no store could possibly have a good enough selection for them because they have read just about all the books in the store.

10.77

a. The Minitab output for testing the independence between age and dependents is given here:
```
Pearson Chi-Square = 62.293, DF = 4, P-Value = 0.000
Likelihood Ratio Chi-Square = 59.602, DF = 4, P-Value = 0.000
```

The Pearson chi-square is 62.293 with df = 4 and p-value < 0.0001. Therefore, there is strong evidence of an association between dependents and age.

b. The Minitab output for testing the independence between opinion and dependents is given here:
```
Pearson Chi-Square = 82.355, DF = 4, P-Value = 0.000
Likelihood Ratio Chi-Square = 83.914, DF = 4, P-Value = 0.000
```

The Pearson chi-square is 82.355 with df = 4 and p-value < 0.0001. Therefore, there is strong evidence of an association between dependents and opinion.

c. The Minitab output for testing the independence between opinion and age for each level of dependents is given here:

Results for Dependents = 0

```
Pearson Chi-Square = 15.029, DF = 16
Likelihood Ratio Chi-Square = 15.934, DF = 16

* WARNING * 1 cells with expected counts less than 1
* WARNING * Chi-Square approximation probably invalid

* NOTE * 19 cells with expected counts less than 5
```

The Pearson chi-square is 15.029 with df = 16 and p-value = 0.523, which would appear to indicate that opinion and age categories are independent for those employees not having their dependents covered by the plan. However, 76% of the cells have expected counts less than 5, which would invalidate the use of the chi-square approximation to the distribution of the Pearson chi-square statistic.

Results for Dependents = 1

```
Pearson Chi-Square = 21.602, DF = 16, P-Value = 0.157
Likelihood Ratio Chi-Square = 23.693, DF = 16, P-Value = 0.096

* NOTE * 8 cells with expected counts less than 5
```

The Pearson chi-square is 21.602 with df = 16 and p-value = 0.157, which would appear to indicate that opinion and age categories are independent for those employees having their dependents covered by the plan. However, 32% of the cells have expected counts less than 5, which would invalidate the use of the chi-square approximation to the distribution of the Pearson chi-square statistic.

Therefore, it appears that opinion and age are associated, but only through each variable's separate association with the dependents variable.

10.79

a.

Rat Group	Odds
Control	0.1111
Low dose	0.1628
High dose	0.2346

b. $OR_{CL} = 0.6824$

$$\ln(OR_{CL}) = -0.3821$$

$$\hat{\sigma}_{\ln(OR)} = \sqrt{\frac{1}{10} + \frac{1}{90} + \frac{1}{14} + \frac{1}{86}} = 0.4406$$

95% CI for log odds: $-0.3821 \pm 1.96(0.4406) = (-1.2457, 0.4815)$
95% CI for odds: $(0.2877, 1.6185)$

c. $OR_{CH} = 0.4736$

$$\ln(OR_{CL}) = -0.7474$$

$$\hat{\sigma}_{\ln(OR)} = \sqrt{\frac{1}{10} + \frac{1}{90} + \frac{1}{19} + \frac{1}{81}} = 0.4196$$

95% CI for log odds: $-0.7474 \pm 1.96(0.4196) = (-1.5698, 0.0750)$
95% CI for odds: $(0.2081, 1.0779)$

d. $OR_{LH} = 0.6939$

$$\ln(OR_{CL}) = -0.3654$$

$$\hat{\sigma}_{\ln(OR)} = \sqrt{\frac{1}{14} + \frac{1}{86} + \frac{1}{19} + \frac{1}{81}} = 0.3848$$

95% CI for log odds: $-0.3654 \pm 1.96(0.3848) = (-1.1196, 0.3888)$
95% CI for odds: $(0.3264, 1.4752)$

e. Because all CI's for the odds ratio include 1, there appears to be no significant impact of the drug on tumor formation.

10.81

a. The results are summarized in the following table, with $\hat{\sigma}_{\hat{\pi}} = \sqrt{\frac{\hat{\pi}(1-\hat{\pi})}{500}}$ and 95% CI $\hat{\pi} \pm 1.96\hat{\sigma}_{\hat{\pi}}$:

Question	$\hat{\pi}$	$\hat{\sigma}_{\hat{\pi}}$	95% CI
Did not explain?	0.254	0.01947	(0.216, 0.292)
Might bother?	0.916	0.0124	(0.892, 0.940)
Did not ask?	0.471	0.02232	(0.427, 0.515)
Drug not changed?	0.877	0.0147	(0.848, 0.906)

b. It would be important to know how the patients were selected, how the questions were phrased, the condition of the illness, and many other factors.

10.83 The combined rate for Anglo-Saxon and German: $\hat{\pi}_1 = \frac{7+6}{55+58} = 0.1150$.

The combined rate for the other four groups: $\hat{\pi}_1 = \frac{34+38+20+31}{52+54+30+49} = 0.6649$.

$$\hat{\sigma}_{\hat{\pi}_1 - \hat{\pi}_2} = \sqrt{\frac{0.1150(0.8850)}{113} + \frac{0.6649(0.3351)}{185}} = 0.0459$$

$H_0: \pi_1 = \pi_2$ versus $H_a: \pi_1 \neq \pi_2$

$z = \dfrac{0.1150 - 0.6649}{0.0459} = -11.98 \Rightarrow$ p-value $< 0.0001 \Rightarrow$ We reject H_0. There is significant evidence of a difference in the rates for the two combined groups.

10.85 $\bar{y} = \sum_i \dfrac{y_i f_i}{500} = \dfrac{0(233) + \cdots + 7(3)}{500} = 1.146$

After combining the last three categories so that all $E_i > 1$ and only one $E_i < 5$, we obtain the following using a Poisson distribution with $\mu = 1.2$:

Mites/leaf (k_i)	0	1	2	3	4	≥ 5
$\pi_i = P(y = k_i)$	0.3012	0.3614	0.2119	0.0867	0.0260	0.0078
$E_i = 40\pi_i$	150.65	180.7	108.5	43.35	13	3.9
n_i	233	127	57	33	30	20

$\chi^2 = \sum_i \dfrac{(n_i - E_i)^2}{E_i} = 176.6$ with df $= 6 - 1 = 5 \Rightarrow$ p-value $< 0.0001 \Rightarrow$ We reject H_0. There is significant evidence that the data do not fit a Poisson distribution with $\mu = 1.2$.

10.87

$\chi^2_{MH} = \dfrac{\left\{\sum_h \left(n_{h11} - \dfrac{n_{h1\bullet} n_{h\bullet 1}}{n_{h\bullet\bullet}}\right)\right\}^2}{\sum_h \dfrac{n_{h1\bullet} n_{h2\bullet} n_{h\bullet 1} n_{h\bullet 2}}{n_{h\bullet\bullet}^2 (n_{h\bullet\bullet} - 1)}}$

Numerator: $\left\{\sum_h \left(n_{h11} - \dfrac{n_{h1\bullet} n_{h\bullet 1}}{n_{h\bullet\bullet}}\right)\right\}^2 = \left\{\left(90 - \dfrac{160(200)}{400}\right) + \left(95 - \dfrac{194(200)}{400}\right) + \left(110 - \dfrac{230(200)}{400}\right)\right\}^2 = 9$

Denominator:

$\sum_h \dfrac{n_{h1\bullet} n_{h2\bullet} n_{h\bullet 1} n_{h\bullet 2}}{n_{h\bullet\bullet}^2 (n_{h\bullet\bullet} - 1)} = \dfrac{(160)(240)(200)(200)}{(400)^2 399} + \dfrac{(95)(99)(200)(200)}{(400)^2 399} + \dfrac{(110)(120)(200)(200)}{(400)^2 399} = 38.2237$

$\chi^2_{MH} = \dfrac{9}{38.2237} = 0.2355$, df = 1, p-value = 0.6275

There is not significant evidence that the number of previous alcohol-related arrests depends on gender.

10.89

$$\chi^2_{MH} = \frac{\left\{\sum_h \left(n_{h11} - \frac{n_{h1\bullet}n_{h\bullet 1}}{n_{h\bullet\bullet}}\right)\right\}^2}{\sum_h \frac{n_{h1\bullet}n_{h2\bullet}n_{h\bullet 1}n_{h\bullet 2}}{n_{h\bullet\bullet}^2(n_{h\bullet\bullet}-1)}}$$

Numerator:

$$\left\{\sum_h \left(n_{h11} - \frac{n_{h1\bullet}n_{h\bullet 1}}{n_{h\bullet\bullet}}\right)\right\}^2 = \left\{\left(38 - \frac{105(258)}{450}\right) + \left(25 - \frac{103(145)}{300}\right) + \left(36 - \frac{131(102)}{300}\right)\right\}^2 = 3082.841$$

Denominator:

$$\sum_h \frac{n_{h1\bullet}n_{h2\bullet}n_{h\bullet 1}n_{h\bullet 2}}{n_{h\bullet\bullet}^2(n_{h\bullet\bullet}-1)} = \frac{(105)(345)(258)(192)}{(450)^2 449} + \frac{(103)(197)(145)(155)}{(300)^2 299} + \frac{(102)(198)(131)(169)}{(300)^2 299} = 53.2982$$

$$\chi^2_{MH} = \frac{3082.841}{53.2982} = 57.841, \text{ df} = 1, \text{ p-value} < 0.0001$$

There is significant evidence that the number of children depends on annual income.

Chapter 11

Linear Regression and Correlation

11.1
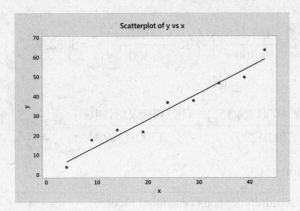

11.3
    a. The following is a scatterplot for the given data values:

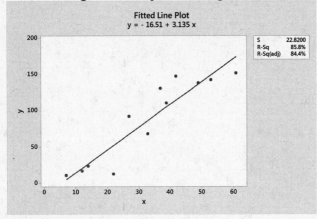

    b. The line should look similar to above.
    c. Answers should be near 110.

11.5
    a. $\hat{y} = -16.51 + 3.135(100) = 296.99$
    b. The prediction is not valid as the x value is well outside the range of the data.

11.7
a. The following is a scatterplot of the data values (with the regression line):

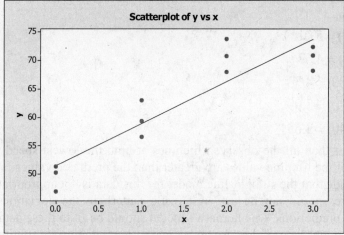

b. $\hat{y} = 51.456 + 7.411x$ See part (a) for the plot with the least-squares line.

c. Firmness does not appear to be constantly increasing. It tapers off, and perhaps decreases slightly, from 2% to 3%.

d. $\hat{y} = 51.456 + 7.411(1.5) = 62.57$

11.9
a. A scatterplot of the data is given here:

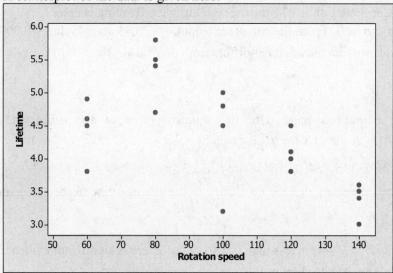

From the plot of the data, there is a definite pattern in the lifetime of the bits as the speed is increased. The lifetimes initially increase, then decrease, as the speed is increased. The relationship between lifetime and speed is not a straight line.

b. The lifetime of one of the bits at speed 100 is quite a bit smaller than the other three lifetimes at this speed. However, because a speed of 100 is the average of all speeds in the range of speeds, this lifetime has very low leverage and hence cannot have high influence.

11.11

a. The prediction equation is $\hat{y} = 6.03 - 0.017x$. Thus, we replace x with the speeds of 60, 80, 100, 120, and 140 to obtain the values:

$x = 60 \Rightarrow \hat{y} = 6.03 - 0.017(60) = 5.01$

$x = 80 \Rightarrow \hat{y} = 6.03 - 0.017(80) = 4.67$

$x = 100 \Rightarrow \hat{y} = 6.03 - 0.017(100) = 4.33$

$x = 120 \Rightarrow \hat{y} = 6.03 - 0.017(120) = 3.99$

$x = 140 \Rightarrow \hat{y} = 6.03 - 0.017(140) = 3.65$

b. The predicted values are larger than all the observed lifetimes at both the lowest speed, 60, and the highest speed, 140. Most of the lifetime values are greater than the predicted values for speeds 80 to 120. Thus we can conclude that the straight-line model for this data is not appropriate. If the fitted line is reasonable, there should not be any systematic pattern in the deviations of the observed data values from the predictions. An alternative model should be fit to these data values.

c. Instead of a transformation, it might be useful to use both x and x^2 as predictors for lifetimes. This would use multiple regression, which is covered in Chapters 12 and 13.

11.13

a. From the computer output on page 625, the standard error of the estimated intercept is $SE(\hat{\beta}_0) = \hat{\sigma}_{\hat{\beta}_0} = 1.953$. A 95% CI for β_0 is given by

$\hat{\beta}_0 \pm t_{0.025,10} SE(\hat{\beta}_0) = 51.456 \pm 2.228(1.953) \Rightarrow (47.10, 55.81)$.

b. The regression line goes through the point (0, 0); that is, the intercept for the regression line is 0. That is, when Pectin Concentration is 0%, the average Firmness Reading is zero.

c. $H_0: \beta_0 = 0$ versus $H_a: \beta_0 \neq 0$. From the computer output, $t = 26.35$ and p-value < 0.0001. Thus, there is significant evidence that the intercept differs from 0.

d. p-value < 0.0001

11.15

a. From the computer output on page 626, the standard error of the estimated slope is $SE(\hat{\beta}_1) = \hat{\sigma}_{\hat{\beta}_1} = 0.216418$. A 95% CI for β_1 is given by

$\hat{\beta}_1 \pm t_{0.025,98} SE(\hat{\beta}_1) \Rightarrow 2.7633 \pm 1.984(0.216418) \Rightarrow (2.33, 3.19)$.

b. As the number of items increases, there is no change in the average time needed to assemble for shipment.

c. $H_a: \beta_1 > 0$

d. The p-value for testing $H_a: \beta_1 > 0$ is less than 0.0001. Thus, there is significant evidence that the slope is greater than 0. The data do support the research hypothesis.

11.17

$H_0: \beta_1 = 0$ versus $H_a: \beta_1 \neq 0$. From the computer output, $t = 12.77$ and p-value < 0.0001. Thus, there is significant evidence that the slope differs from 0.

11.19 Minitab output is given here:
Regression Analysis: log10 Recovery versus Time (minutes)

```
The regression equation is
log10 Recovery = 1.67 - 0.0159 Time (minutes)


Predictor            Coef   SE Coef       T      P
Constant          1.67243   0.05837   28.65  0.000
Time (minutes)  -0.015914  0.001651   -9.64  0.000


S = 0.111354   R-Sq = 89.4%   R-Sq(adj) = 88.5%


Analysis of Variance

Source          DF      SS      MS       F      P
Regression       1  1.1523  1.1523   92.93  0.000
Residual Error  11  0.1364  0.0124
Total           12  1.2887
```

a. $\hat{y} = 1.67 - 0.0159x$
b. $s_\varepsilon = 0.1114$
c. $SE(\hat{\beta}_0) = 0.05837 \quad SE(\hat{\beta}_1) = 0.001651$

11.21 $1.67243 \pm 2.201(0.05837) \Rightarrow (1.54, 1.80)$

We are 95% confident that the average value of the logarithm of biological recovery percentage at time zero is between 1.54 and 1.80.

11.23

a. $\hat{\sigma}_\varepsilon^2 = (2.10171)^2 = 4.42 \ (= MS(\text{Residual Error}))$
b. $SE(\hat{\beta}_1) = 0.3462$
c. $-1.8673 \pm 2.101(0.3462) \Rightarrow (-2.595, -1.140)$
d. $H_0: \beta_1 = 0$ versus $H_a: \beta_1 \neq 0$. From the computer output, $t = -5.39$ and p-value < 0.0001 < 0.05. Thus, we reject H_0 and conclude there is significant evidence that there is a linear relationship between the amount of time needed to run a 10-km race and the time to exhaustion on a treadmill.

11.25

a. $\hat{y} = 52.819 - 0.9834x$
b. $\hat{\sigma}_\varepsilon^2 = (2.69935)^2 = 7.29 \ (= MS(\text{Residual Error}))$
c. $SE(\hat{\beta}_1) = 0.1083$
d. $-0.9834 \pm 2.101(0.1083) \Rightarrow (-1.211, -0.756)$
e. $H_0: \beta_1 = 0$ versus $H_a: \beta_1 \neq 0$. From the computer output, $t = -9.08$ and p-value < 0.0001 < 0.05. Thus, we reject H_0 and conclude there is significant evidence that there is a linear relationship between yield per acre and the absolute deviation from the ideal planting date.

11.27
a. One run has a size of 20,000 stickers, which is quite high in terms of the other run sizes. This point may be a leverage point. In addition, the residual plot does not appear to have random scatter. The regression assumptions appear to be violated.
b. $\hat{y} = 99.777 + 51.9179x$
c. $S = \sqrt{MSE} = \sqrt{149} = 12.2065$
d. $51.9179 \pm 2.048(0.5865) \Rightarrow (50.718, 53.119)$
e. Intercept: When we run zero stickers, the direct cost is expected to be \$99.78. This is meaningless, since we should have no cost associated with a run of zero stickers.
Slope: For each additional thousand stickers run, we expect the direct cost to increase by \$51.92.

11.29
a. $F = 7837.26$, p-value < 0.0001
b. The p-values are the same. This relationship should hold because the F statistic is the square of the t statistic for testing $H_a : \beta_1 \neq 0$. In this case,
$F = 7837.26 = (88.53)^2 = t^2$ (slight difference due to rounding).

11.31
a. $\hat{y}_{n+1} \pm t_{0.025, 11} s_\varepsilon \sqrt{1 + \dfrac{1}{n} + \dfrac{(30 - \bar{x})^2}{S_{xx}}} \Rightarrow$

$1.195 \pm 2.201(0.1114)\sqrt{1 + \dfrac{1}{13} + \dfrac{(30-30)^2}{4550}} \Rightarrow 1.195 \pm 0.254 \Rightarrow (0.941, 1.449)$

b. The prediction interval in part (a) is wider than the confidence interval for the mean in Exercise 11.33.
c. 95% confidence interval for the mean: We are 95% confident that the mean log biological recovery percentage at 30 minutes will be between 1.127 and 1.263. (Or, we are 95% confident that the mean biological recovery percentage at 30 minutes will be between 13.4% and 18.3%.)
95% prediction interval: We are 95% confident that the log biological recovery percentage for a single sample at 30 minutes will be between 0.941 and 1.449. (Or, we are 95% confident that the biological recovery percentage for a single sample at 30 minutes will be between 8.7% and 28.1%.)

11.33
a. 95% confidence intervals for $E(y)$ at selected values of x:
$x = 4 \Rightarrow (2.6679, 4.3987)$
$x = 5 \Rightarrow (4.2835, 5.4165)$
$x = 6 \Rightarrow (5.6001, 6.7332)$
$x = 7 \Rightarrow (6.6179, 8.3487)$
b. 95% prediction intervals for y at selected values of x:
$x = 4 \Rightarrow (1.5437, 5.5229)$
$x = 5 \Rightarrow (2.9710, 6.7290)$
$x = 6 \Rightarrow (4.2877, 8.0456)$
$x = 7 \Rightarrow (5.4937, 9.4729)$

c. The confidence intervals in part (a) are interpreted as "We are 95% confident that the average weight loss over many samples of the compound when exposed for 4 hours will be between 2.67 and 4.40 pounds." Similar statements can be made for the other hours of exposure.

The prediction intervals in part (b) are interpreted as "We are 95% confident that the weight loss for a single sample of the compound when exposed for 4 hours will be between 1.54 and 5.52 pounds." Similar statements can be made for the other hours of exposure.

11.35
  a. The point estimate for the prediction is the same as in Exercise 11.38: 203.613. The 95% PI for a single observation is given in the output as (178.169, 229.057).
  b. Yes, because $250 does not fall within the 95% prediction interval. In fact, it is considerably higher than the upper value of $229.06.

11.37
  a. The point estimate for the prediction is the same as in Exercise 11.41: 38.2757.
  b. 95% prediction interval: (33.661, 42.889)
  c. The prediction interval is wider than the confidence interval on the mean.
  95% confidence interval for the mean: We are 95% confident that the mean time to run 10km for athletes having a treadmill time of 11 minutes will be between 36.935 and 39.615 minutes.
  95% prediction interval: We are 95% confident that the time to run 10km for an athlete having a treadmill time of 11 minutes will be between 36.935 and 39.615 minutes. The prediction interval is wider because there is more variation for an individual than for a mean, and the confidence interval is for a mean.

11.39
  a. A test of lack of fit will be conducted:
  $$SSP_{exp} = \sum_{ij}(y_{ij} - \bar{y}_{i.})^2 = (28.1 - 27.85)^2 + (27.6 - 27.85)^2 + (32.3 - 32.75)^2 +$$
  $$(33.2 - 32.75)^2 + (34.8 - 34.90)^2 + (35.0 - 34.90)^2 + (38.2 - 38.80)^2 + (39.4 - 38.80)^2 + \text{From the}$$
  $$(43.5 - 45.15)^2 + (46.8 - 45.15)^2 = 6.715$$
  output in Exercise 11.43, SS(Residuals) = 15.6245. Thus,
  $$SSP_{lack} = SS(Residuals) - SS_{exp} = 15.6245 - 6.715 = 8.9095$$
  $$df_{lack} = n - 2 - \sum_i (n_i - 1) = 10 - 2 - 5(2 - 1) = 3$$
  $$df_{exp} = \sum_i (n_i - 1) = 5(2 - 1) = 5$$
  $$F = \frac{8.9095/3}{6.715/5} = 2.21 < 5.41 = F_{0.05, 3, 5}$$
  There is not sufficient evidence that the linear model is inadequate.
  b. $S_{xx} = \sum_i (x_i - \bar{x})^2 = 20$
  95% prediction interval for y at x:
  $$\hat{y} \pm t_{0.025, 8} s_\varepsilon \sqrt{1 + \frac{1}{n} + \frac{(x - \bar{x})^2}{S_{xx}}} = 3.37 + 4.065x \pm 2.306(1.398)\sqrt{1 + \frac{1}{10} + \frac{(x - 8)^2}{20}}$$

For $x = 6, 7, 8, 9,$ and 10 we have

x	$\hat{y}$	95% PI
6	27.760	(24.086, 31.434)
7	31.825	(28.369, 35.281)
8	35.890	(32.510, 39.270)
9	39.955	(36.499, 43.411)
10	44.020	(40.346, 47.694)

11.41 A test of lack of fit will be conducted:

$SSP_{exp} = \sum_{ij}(y_{ij}-\bar{y}_{i\cdot})^2 = (5-5)^2 + (7-5)^2 + (3-5)^2 + (10-12)^2 + (14-12)^2 +$

$(15-16)^2 + (17-16)^2 + (20-20)^2 + (21-20)^2 + (19-20)^2 + (23-26)^2 +$

$(29-26)^2 + (28-29.667)^2 + (31-29.667)^2 + (30-29.667)^2$

$= 42.667 =$ SSE from output on page 637

From the output on page 638, SS(Residuals) = 284.95. Thus,

$SSP_{lack} =$ SS(Residuals) $- SS_{exp} = 284.95 - 42.667 = 242.283$

$df_{lack} = n - 2 - \sum_i (n_i - 1) = 15 - 2 - (2+1+1+2+1+2) = 4$

$df_{exp} = \sum_i (n_i - 1) = 2+1+1+2+1+2 = 9$

$F = \dfrac{242.283/4}{42.667/9} = 12.78 > 3.63 = F_{0.05,4,9}$

There is sufficient evidence that the linear model is inadequate.

11.43

a. $r_{yx}^2 = \dfrac{SS(Total) - SS(Residual)}{SS(Total)} = \dfrac{1{,}171{,}919 - 4172}{1{,}171{,}919} = 0.9964$. The output gives R-Sq = 0.996, which is the same (to rounding) as the value we calculated.

b. Since β_1 is positive, there must be a general positive relationship between the values of y and x. Thus $r_{yx} = +\sqrt{0.9964} = 0.9982$.

c. In general, if the relationship between the values of y and x are fairly consistent across the range of values for x, then a wider range of values for x yields a larger value for r_{yx}. Examining the plot of Exercise 11.30 confirms the consistency of the relationship. Thus, we would expect r_{yx} to be smaller for the restricted data set.

11.45

a. $r = 0.695$ The correlation is moderate to strong and positive which is in agreement with the plot.
b. $\rho = 0.704$
c. The Pearson correlation is more influenced by the outlier.

11.47
a. This data value has high leverage and high influence (outlier in the x and y direction) although the influence is greater.
b. Slope would increase if the point was removed.
c. The residual standard deviation would decrease if the point was removed.
d. The correlation would increase if the point was removed.

11.49
a. $r_{yx} = 0.995$

b. $z = \dfrac{1}{2}\ln\left(\dfrac{1+r_{yx}}{1-r_{yx}}\right) = \dfrac{1}{2}\ln\left(\dfrac{1+0.995}{1-0.995}\right) = 2.9945$

$z_1 = z - \dfrac{z_{\alpha/2}}{\sqrt{n-3}} = 2.9945 - \dfrac{1.96}{\sqrt{15-3}} = 2.42870$

$z_2 = z + \dfrac{z_{\alpha/2}}{\sqrt{n-3}} = 2.9945 + \dfrac{1.96}{\sqrt{15-3}} = 3.56030$

$\left(\dfrac{e^{2z_1}-1}{e^{2z_1}+1}, \dfrac{e^{2z_2}-1}{e^{2z_2}+1}\right) = \left(\dfrac{e^{2(2.42870)}-1}{e^{2(2.42870)}+1}, \dfrac{e^{2(3.56030)}-1}{e^{2(3.56030)}+1}\right) = (0.985, 0.998)$

With 95% confidence we would estimate that the correlation coefficient is between 0.985 and 0.998.

c. A very large positive value for the correlation coefficient indicates that as the price at one garage increases (decreases), the price at the other garage tends to increase (decrease) as well. This does not, however, mean that the two garages are providing nearly identical estimates for the repairs.

11.51
a. The following table gives the intermediate calculations and 95% confidence intervals for the six correlations:

	r	z	z1	z2	LCL	UCL
MaleV/FemaleV	0.708	0.883162	0.531135894	1.235189	0.486249	0.844078
MaleV/MaleM	-0.133	-0.13379	-0.48581903	0.218234	-0.45089	0.214834
MaleV/FemaleM	-0.288	-0.29638	-0.64841039	0.055642	-0.5706	0.055585
FemaleV/MaleM	0.392	0.414161	0.062134587	0.766187	0.062055	0.644707
FemaleV/FemaleM	0.264	0.270403	-0.08162316	0.62243	-0.08144	0.552817
MaleM/FemaleM	0.977	2.226921	1.874894412	2.578947	0.954036	0.988558

b. The correlations between verbal scores for male and female students are strong and differ from zero. The correlations between math scores for male and female students are strong and differ from zero. The correlations between math scores for male and verbal scores female students are significantly different from zero. The other three correlations do not differ from zero, since zero is included in each of those four confidence intervals.

c. The results in part (b) match our answer to part (c) in Exercise 11.58.

11.53

a. Standard/Old: $\hat{y} = 15.432 + 2.7897(20) = 71.226$

   New: $\hat{y} = 12.642 + 2.7168(20) = 66.978$

b. Standard/Old, 95% CI:

$$71.226 \pm 2.101(5.96512)\sqrt{\frac{1}{20} + \frac{(20-28.49)^2}{542.948}} = (65.868, 76.584)$$

   New, 95% CI:

$$66.978 \pm 2.101(6.12441)\sqrt{\frac{1}{20} + \frac{(20-28.21)^2}{535.7335}} = (61.583, 72.373)$$

c. Since the two intervals overlap, it does not appear that the average gas consumption has been reduced by using the new form of insulation.

d. Note that 50 is well beyond the observed data and hence predictions should not be made due to extreme extrapolation.

11.55

a. Slope = 1.804 and intercept = 47.1, yielding $\hat{y} = 47.15 + 1.80x$ for the least-squares line.

b. The estimated slope is approximately 1.80. This implies that as yearly income increases $1000, the average sale price of the home increases $1800. The estimated intercept would be the average sale price of homes bought by someone with 0 yearly income. Because the data set did not contain any data values with yearly incomes close to 0, the intercept does not have a meaningful interpretation.

c. The residual standard deviation is "Root MSE" = 14.482.

11.57

a. Yes. The data values fall approximately along a straight line.

b. $\hat{y} = 12.51 + 35.83x$

11.59

a. No, because the interpretation for β_0 is the average weight gain for chickens who did not ingest lysine in their diet. However, lysine was placed in all the feed. Thus, all chickens consumed some lysine. There were no observations close to $x = 0$.

b. The model $y = \beta_1 x + \varepsilon$ forces the estimated regression line to pass through the origin, the point (0, 0). The model $y = \beta_0 + \beta_1 x + \varepsilon$ does not necessarily force the fitted line to pass through the origin. This provides greater flexibility in fitting models where there was no data collected near $x = 0$.

11.61
a. The scatterplot is given below.

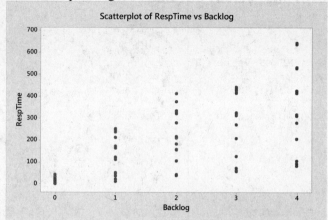

While there is an increasing relationship, I'd be skeptical of fitting a linear model to the data as the variability is increasing as backlog (x) increases.

b. Regression output below

Regression Analysis: RespTime versus Backlog

```
Analysis of Variance

Source          DF   Adj SS    Adj MS   F-Value   P-Value
Regression       1   1016652   1016652    67.81     0.000
  Backlog        1   1016652   1016652    67.81     0.000
Error           73   1094454    14993
  Lack-of-Fit    3    20577     6859      0.45      0.720
  Pure Error    70   1073877    15341
Total           74   2111106


Model Summary

      S    R-sq   R-sq(adj)   R-sq(pred)
122.444   48.16%    47.45%      45.14%


Coefficients

Term       Coef   SE Coef   T-Value   P-Value   VIF
Constant   39.3     24.5      1.60     0.113
Backlog    82.3     10.0      8.23     0.000    1.00


Regression Equation

RespTime = 39.3 + 82.3 Backlog
```

c. Regression output below.

Regression Analysis: log(RespTime) versus Backlog

```
Analysis of Variance

Source         DF   Adj SS    Adj MS   F-Value  P-Value
Regression      1   14.668   14.6683    83.92    0.000
  Backlog       1   14.668   14.6683    83.92    0.000
Error          73   12.759    0.1748
  Lack-of-Fit   3    3.960    1.3200    10.50    0.000
  Pure Error   70    8.799    0.1257
Total          74   27.427


Model Summary

      S      R-sq   R-sq(adj)   R-sq(pred)
0.418069    53.48%    52.84%       50.79%


Coefficients

Term        Coef    SE Coef   T-Value   P-Value   VIF
Constant   1.4153   0.0836    16.93     0.000
Backlog    0.3127   0.0341     9.16     0.000    1.00


Regression Equation

log(RespTime) = 1.4153 + 0.3127 Backlog
```

d. The lack-of-fit test suggest that the model based on log(RespTime) is more appropriate.

11.63

```
Analysis of Variance

Source         DF    Adj SS    Adj MS   F-Value  P-Value
Regression      1   1016652   1016652    67.81    0.000
  Backlog       1   1016652   1016652    67.81    0.000
Error          73   1094454     14993
  Lack-of-Fit   3     20577      6859     0.45    0.720
  Pure Error   70   1073877     15341
Total          74   2111106
```

a. The p-value for 'lack-of-fit' is 0.720 which leads to a conclusion that the linear model is appropriate.

Regression Analysis: log(RespTime) versus Backlog

```
Analysis of Variance

Source         DF   Adj SS    Adj MS   F-Value  P-Value
Regression      1   14.668   14.6683    83.92    0.000
  Backlog       1   14.668   14.6683    83.92    0.000
Error          73   12.759    0.1748
  Lack-of-Fit   3    3.960    1.3200    10.50    0.000
  Pure Error   70    8.799    0.1257
Total          74   27.427
```

b. The p-value for 'lack-of-fit' is 0.000 which leads to a conclusion that the linear model is not appropriate.

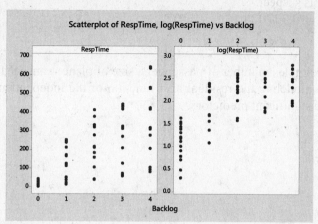

c. The graphs agree as the plot of the log(resptime) vs. backlog shows a bit of curvature.

11.65

a. The plot shows a pretty clear linear relationship.

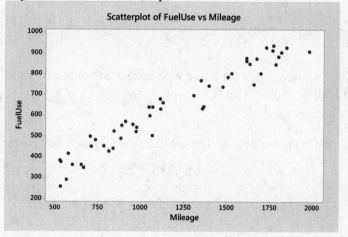

b. The regression output is shown below.
```
Coefficients

Term         Coef    SE Coef   T-Value   P-Value    VIF
Constant    123.9      19.8      6.26     0.000
Mileage    0.4319    0.0157     27.48     0.000    1.00


Regression Equation

FuelUse = 123.9 + 0.4319 Mileage
```

c. For every additional hundred miles of flight, an additional 0.4319 gallons of fuel is expected to be consumed.

Chapter 11: Linear Regression and Correlation

d. It would be suspect to interpret the intercept in this case because it would be an extrapolation. A plausible explanation would be the amount of fuel used for taxi, takeoff, and landing, but again, the interpretation is suspect.

e. $r_{yx} = 0.968$; $R^2 = 0.937024$

11.67
a. Answers vary. Examples: number of passengers, size of plane, transoceanic, etc.
b. Run a regression modeling the residual as a function of the independent variables and see if the new variables are significant predictors.

11.69

Prediction for ActVol

```
Regression Equation

ActVol = -0.95 + 1.1907 EstVol


Variable   Setting
EstVol        13


   Fit      SE Fit          95% CI              95% PI
14.5315   0.318490   (13.8727, 15.1904)   (12.1306, 16.9325)
```

a. The predicted actual volume for a tree with estimated volume of 13 ft^3 is 14.5315 ft^3
b. 95% PI: (12.1306, 16.9325) (in ft^3)

11.71
a. The scatter plot is shown below. The curvature is lessened, but the increasing variance remains.

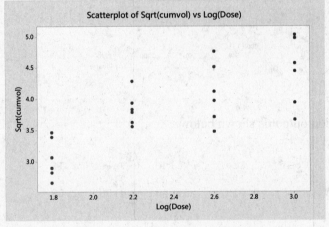

b. The regression output is shown below.

Regression Analysis: Sqrt(cumvol) versus Log(Dose)

```
Analysis of Variance

Source        DF  Adj SS  Adj MS  F-Value  P-Value
Regression     1  5.8734  5.8734   33.10    0.000
  Log(Dose)    1  5.8734  5.8734   33.10    0.000
Error         22  3.9042  0.1775
  Lack-of-Fit  3  0.4334  0.1445    0.79    0.514
  Pure Error  19  3.4709  0.1827
Total         23  9.7777


Model Summary

       S    R-sq  R-sq(adj)  R-sq(pred)
0.421267  60.07%     58.25%      51.94%


Coefficients

Term        Coef  SE Coef  T-Value  P-Value   VIF
Constant   1.239    0.463     2.67    0.014
Log(Dose)  1.091    0.190     5.75    0.000  1.00


Regression Equation

Sqrt(cumvol) = 1.239 + 1.091 Log(Dose)


Fits and Diagnostics for Unusual Observations

Obs  Sqrt(cumvol)    Fit   Resid  Std Resid
 22         3.674  4.522  -0.847      -2.14  R

R  Large residual
```

$sqrt(cumvol) = 1.24 + 1.09\ln(dose)$

c. The test for lack-of-fit fails to reject ($p = 0.514$) which shows there is not sufficient evidence the model doesn't fit. The p-value increased with the transformed data which implies the fit has improved somewhat.

d. 95% CI when ln(dose)=2.70805: (15.829,19.447)

e. Using the transformed data, the CI has shifted slightly to larger values.

11.73

Regression Analysis: Rate versus Mileage

```
Model Summary

      S    R-sq  R-sq(adj)  R-sq(pred)
2.04204  98.78%     98.76%      98.57%


Coefficients

Term          Coef   SE Coef  T-Value  P-Value   VIF
Constant     9.793     0.439    22.30    0.000
Mileage   0.050323  0.000823    61.14    0.000  1.00


Regression Equation

Rate = 9.793 + 0.050323 Mileage
```

a. $\hat{y} = 9.793 + 0.0503x$; $s = 2.04204$
b. 90% CI: $\hat{\beta}_1 \pm t_{0.05,46} SE(\hat{\beta}_1) \Rightarrow 0.0503 \pm 1.679(0.000823) \Rightarrow (0.0489, 0.0516)$

11.75
Prediction for Rate

```
Regression Equation

Rate = 9.793 + 0.050323 Mileage


Variable  Setting
Mileage       350


    Fit   SE Fit         95% CI              95% PI
27.4062  0.297104  (26.8081, 28.0042)  (23.2525, 31.5599)
```

The 95% PI when $x = 350$ is (23.2525, 31.5599). The extrapolation problem isn't an issue here.

11.77 There is a single point in the upper right hand corner of the plot that is a considerable distance from the rest of the data points. A point located in this region will cause the fitted line to be drawn away from the line that would be fitted to the data values with this point excluded. Thus, the regression line given in the plot seems to be somewhat misfitted. A regression line fitted to the data set excluding this point would most likely yield a line with a negative slope.

11.79
  a. $Expend = 180.38 - 1.353(TownPop)$
  b. Without outlier: Prediction townpop=40 → 126.6
     With outlier: Prediction townpop=40 → 143.92
  c. There is a marked difference in the estimates of the line with and without the outlier included which heavily affects all predictions. I would feel much more comfortable using the data without the outlier as those points are more indicative of the population of interest.

Chapter 11: Linear Regression and Correlation

11.81 The Minitab output is given here:
Regression Analysis: Yield versus Nitrogen

```
The regression equation is
Yield = 11.8 + 6.50 Nitrogen

Predictor      Coef    SE Coef     T       P
Constant     11.778     1.887    6.24   0.000
Nitrogen     6.5000     0.8736   7.44   0.000

S = 2.13995    R-Sq = 88.8%   R-Sq(adj) = 87.2%

Analysis of Variance

Source           DF       SS       MS       F       P
Regression        1   253.50   253.50   55.36   0.000
Residual Error    7    32.06     4.58
  Lack of Fit     1     2.72     2.72    0.56   0.484
  Pure Error      6    29.33     4.89
Total             8   285.56
```

$\hat{y} = 11.778 + 6.5x$

From the output, $MS_{Lack} = 2.72$, $MS_{exp} = 4.89 \Rightarrow F = 2.72/4.89 = 0.56$ with df = 1, 6. This yields p-value = 0.484 which implies there is no indication of lack of fit in the model.

11.83
a. A scatterplot of the data is given here:

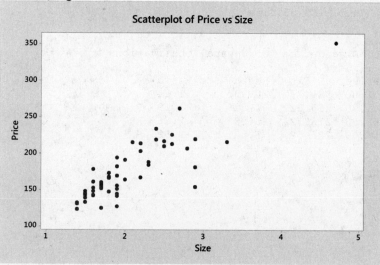

b. One of the houses in the study is shown as having a size of 4.7. It appears to be an outlier, with a value well outside the range of sizes that may be fitted to the data. Because the point has an x-value far from $\bar{x}$, the point has high leverage.

Chapter 11: Linear Regression and Correlation

c. The Minitab output is given here:

Regression Analysis: Price versus Size

```
Analysis of Variance

Source         DF   Adj SS   Adj MS   F-Value   P-Value
Regression      1    66577  66576.6    152.12     0.000
  Size          1    66577  66576.6    152.12     0.000
Error          55    24071    437.7
  Lack-of-Fit  16    14101    881.3      3.45     0.001
  Pure Error   39     9970    255.6
Total          56    90648


Model Summary

     S      R-sq   R-sq(adj)   R-sq(pred)
20.9204    73.45%     72.96%       70.46%


Coefficients

Term        Coef   SE Coef   T-Value   P-Value   VIF
Constant    54.3      10.0      5.42     0.000
Size       59.03      4.79     12.33     0.000   1.00


Regression Equation

Price = 54.3 + 59.03 Size


Fits and Diagnostics for Unusual Observations

Obs    Price      Fit    Resid   Std Resid
  4   352.00   331.77    20.23        1.24   X
 13   181.00   225.52   -44.52       -2.19   R
 26   262.00   213.72    48.28        2.36   R
 47   154.00   225.52   -71.52       -3.52   R

R  Large residual
X  Unusual X
```

d. Minitab output for the data set with the outlier removed is given here:

Regression Analysis: Price versus Size

```
Method

Rows unused  1

Analysis of Variance

Source          DF   Adj SS   Adj MS   F-Value   P-Value
Regression       1    34723  34723.1     80.15     0.000
  Size           1    34723  34723.1     80.15     0.000
Error           54    23395    433.2
  Lack-of-Fit   15    13425    895.0      3.50     0.001
  Pure Error    39     9970    255.6
Total           55    58118

Model Summary

      S    R-sq   R-sq(adj)  R-sq(pred)
20.8144  59.75%      59.00%      55.03%

Coefficients

Term       Coef   SE Coef   T-Value   P-Value    VIF
Constant   63.2      12.3      5.16     0.000
Size      54.33      6.07      8.95     0.000   1.00

Regression Equation

Price = 63.2 + 54.33 Size

Fits and Diagnostics for Unusual Observations

Obs   Price     Fit    Resid   Std Resid
  8  216.00  242.50   -26.50       -1.40   X
 13  181.00  220.77   -39.77       -2.01   R
 26  262.00  209.91    52.09        2.59   R
 47  154.00  220.77   -66.77       -3.37   R

R  Large residual
X  Unusual X
```

Note that the slope has changed quite a bit (59.03 to 54.33) since the outlier was of high leverage.

e. The residual standard deviations are $s_\varepsilon = 20.92$ with the outlier and $s_\varepsilon = 20.81$ without the outlier. There hasn't been a considerable reduction in the residual standard deviation because while the outlier had high leverage, it was along the line of the data.

11.85
a. From the Minitab output, a 95% PI for the selling price when the size is 5 is (278.865, 390.845). The prediction may be somewhat suspect since a house of size 5 (5000 square feet) is beyond all the data values in the study.
b. A scatterplot of the data is given here:

The variance in selling price appears to increase as the size of the houses increases.
c. The PI may not be valid since the statistical procedures depend on the condition that the variance remains constant across the range of x-values in addition to the extrapolation problem.

11.87
a. $t = -6.63 \Rightarrow$ p-value $= P(|t_{30}| \geq 6.63) < 0.0001$. Thus, there is very strong evidence that density is related to sales.
b. 95% CI for β_1: $-12.893 \pm 2.042(1.946) \Rightarrow (-16.867, -8.919)$

11.89 Minitab output is given here:
Regression Analysis: Sales versus 1/Density

```
The regression equation is
Sales = 27.7 + 199 1/Density

Predictor      Coef   SE Coef      T      P
Constant     27.704     3.945   7.02  0.000
1/Density    199.22     11.74  16.97  0.000

S = 10.4802   R-Sq = 90.6%   R-Sq(adj) = 90.3%
```

a. The density variable is number of homes per acre. Therefore, 1/density would be the number of acres per home, that is, the average lot size of a home. If 1/density equals 0.5, then the average lot size in this zip code area would be a 0.5-acre lot.

b. A scatterplot of the data is given here:

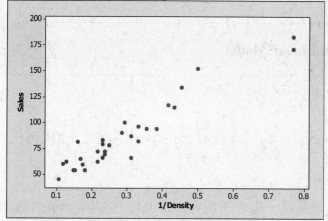

The plotted points appear to fit a straight-line model reasonably well.

c. The correlation is obtained by taking the square root of R^2 and giving it the same sign as the estimated slope. $r = +\sqrt{0.906} = 0.9518$. The magnitude of r has increased (0.7707 versus 0.9518). The relationship between sales and lot size (1/density) is more linear than the relationship between sales and density. This results in an increase in the magnitude of the correlation coefficient.

11.91 For testing $H_0: \beta_1 = 0$ versus $H_a: \beta_1 \neq 0$, $t = 3.89$ with p-value < 0.001. The p-value is less than $\alpha = 0.01$. Thus, we reject H_0 and conclude there is significant evidence that the slope is different from 0.

11.93 Some of the important points to mention are:
- A correlation of 0.801 indicates that the linear relationship between the estimated cost and the actual cost is quite strong. This does not mean that the estimated cost itself is accurate. For instance, the estimated cost could always be off by a factor of 3 (i.e., consistently actual cost = 3 × estimated cost). This would result in a perfect correlation between estimated cost and actual cost, but the estimated cost would always be considerably different from the actual cost.
- The p-value is low, which corresponds to the high correlation. The low p-value affirms the strong linear relationship between estimated cost and actual cost.
- The estimate slope is 1.25. This implies that the actual cost is generally higher than the estimated cost. Thus, the estimating method may not be a reasonable method.
- There are two observations that might need examination. The point with estimated cost = 186,200 and actual cost 152,134 is very much below the general pattern in the data. The second point would be estimated cost = 178,300 and actual cost = 215,200, which is somewhat above the general pattern in the data.

Chapter 12

Multiple Regression and the General Linear Model

12.1
a. $MPG = \beta_0 + \beta_1 c_i + \beta_2 v_i + \beta_3 w_i + \varepsilon_i$
b. $MPG = \beta_0 + \beta_1 c_i + \beta_2 v_i + \beta_3 w_i + \beta_4 c_i^2 + \beta_5 v_i^2 + \beta_6 w_i^2 + \beta_7 c_i v_i + \beta_8 v_i w_i + \beta_9 c_i w_i + \varepsilon_i$

12.3
a. The model is shown below.
$$MPG = \beta_0 + \beta_1 c_i + \beta_2 v_i + \beta_3 w_i + \beta_4 c_i^2 + \beta_5 v_i^2 + \beta_6 w_i^2 + \beta_7 c_i v_i + \beta_8 v_i w_i + \beta_9 c_i w_i +$$
$$\beta_{10} t_{1i} + \beta_{11} c_i t_{1i} + \beta_{12} v_i t_{1i} + \beta_{13} w_i t_{1i} + \beta_{14} c_i^2 t_{1i} + \beta_{15} v_i^2 t_{1i} + \beta_{16} w_i^2 t_{1i} + \beta_{17} c_i v_i t_{1i} + \beta_{18} v_i w_i t_{1i}$$
$$+ \beta_{19} c_i w_i t_{1i} + \beta_{20} t_{2i} + \beta_{21} c_i t_{2i} + \beta_{22} v_i t_{2i} + \beta_{23} w_i t_{2i} + \beta_{24} c_i^2 t_{2i} + \beta_{25} v_i^2 t_{2i}$$
$$+ \beta_{26} w_i^2 t_{2i} + \beta_{27} c_i v_i t_{2i} + \beta_{28} v_i w_i t_{2i} + \beta_{29} c_i w_i t_{2i} + \varepsilon_i$$

$$t_{1i} = \begin{cases} 1 & \text{if rear} - \text{wheel} \\ 0 & \text{Otherwise} \end{cases}; \quad t_{2i} = \begin{cases} 1 & \text{if front} - \text{wheel} \\ 0 & \text{Otherwise} \end{cases}$$

b. The following table gives the slopes and intercepts associated with the different varieties:

Drive	Slopes
Rear-wheel	$MPG = \beta_0 + (\beta_1 + \beta_{11})c_i + (\beta_2 + \beta_{12})v_i + (\beta_3 + \beta_{13})w_i + (\beta_4 + \beta_{14})c_i^2 + (\beta_5 + \beta_{15})v_i^2 + (\beta_6 + \beta_{16})w_i^2$ $+ (\beta_7 + \beta_{17})c_i v_i + (\beta_8 + \beta_{18})v_i w_i + (\beta_9 + \beta_{19})c_i w_i + \varepsilon_i$
Front-wheel	$MPG = \beta_0 + (\beta_1 + \beta_{21})c_i + (\beta_2 + \beta_{22})v_i + (\beta_3 + \beta_{23})w_i + (\beta_4 + \beta_{24})c_i^2 + (\beta_5 + \beta_{25})v_i^2 + (\beta_6 + \beta_{26})w_i^2$ $+ (\beta_7 + \beta_{27})c_i v_i + (\beta_8 + \beta_{28})v_i w_i + (\beta_9 + \beta_{29})c_i w_i + \varepsilon_i$
All-wheel	$MPG = \beta_0 + \beta_1 c_i + \beta_2 v_i + \beta_3 w_i + \beta_4 c_i^2 + \beta_5 v_i^2 + \beta_6 w_i^2 + \beta_7 c_i v_i + \beta_8 v_i w_i + \beta_9 c_i w_i + \varepsilon_i$

12.5
a. $HHI = \beta_0 + \beta_1 A + \beta_2 BMI + \beta_3 E + \beta_4 SB + \beta_5 M + \beta_6 A * M + \beta_7 BMI * M + \beta_8 E * M + \beta_9 SB * M + \beta_{10} D + \beta_{11} A * D + \beta_{12} BMI * D + \beta_{13} E * D + \beta_{14} SB * D + \varepsilon_i$

$$M = \begin{cases} 1 & \text{if male} \\ 0 & \text{Otherwise} \end{cases}; \quad D = \begin{cases} 1 & \text{if diabetic} \\ 0 & \text{Otherwise} \end{cases}$$

b. The table is shown below

Sex	Diabetic	Model
M	Y	$HHI = (\beta_0 + \beta_5 + \beta_{10}) + (\beta_1 + \beta_6 + \beta_{11})A + (\beta_2 + \beta_7 + \beta_{12})BMI + (\beta_3 + \beta_8 + \beta_{11})E + (\beta_4 + \beta_9 + \beta_{12})SB + \varepsilon_i$
M	N	$HHI = (\beta_0 + \beta_5) + (\beta_1 + \beta_6)A + (\beta_2 + \beta_7)BMI + (\beta_3 + \beta_8)E + (\beta_4 + \beta_9)SB + \varepsilon_i$
F	Y	$HHI = (\beta_0 + \beta_{10}) + (\beta_1 + \beta_{11})A + (\beta_2 + \beta_{12})BMI + (\beta_3 + \beta_{11})E + (\beta_4 + \beta_{12})SB + \varepsilon_i$
F	N	$HHI = \beta_0 + \beta_1 A + \beta_2 BMI + \beta_3 E + \beta_4 SB + \varepsilon_i$

12.7
  a. Means for boys and girls enrolled in the new program where English is spoken in the home.

Gender	Mean
Boy	$\beta_0 + \beta_1 + \beta_2 + \beta_4$
Girl	$\beta_0 + \beta_1 + \beta_2 + \beta_3 + \beta_4 + \beta_5 + \beta_6$

  b. Means for boys and girls enrolled in the new program where English is not spoken in the home. Only girls requested.

Gender	Mean
Boy	$\beta_0 + \beta_1$
Girl	$\beta_0 + \beta_1 + \beta_3 + \beta_5$

  c. Difference in means for boys and girls enrolled in the new program based on English spoken in the home. Only boys requested.

Gender	(English) - (non-English)
Boy	$\beta_2 + \beta_4$
Girl	$\beta_2 + \beta_4 + \beta_6$

12.9

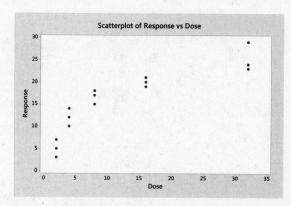

  a.

  b. LINEAR MODEL

Regression Analysis: Response versus Dose

```
Analysis of Variance

Source        DF   Adj SS   Adj MS   F-Value   P-Value
Regression     1   590.92   590.916    44.28     0.000
  Dose         1   590.92   590.916    44.28     0.000
Error         13   173.48    13.345
  Lack-of-Fit  3   130.15    43.384    10.01     0.002
  Pure Error  10    43.33     4.333
Total         14   764.40


Model Summary

      S    R-sq   R-sq(adj)   R-sq(pred)
3.65307   77.30%     75.56%       68.86%
```

```
Coefficients

Term        Coef    SE Coef   T-Value   P-Value   VIF
Constant    8.67    1.43      6.07      0.000
Dose        0.5753  0.0865    6.65      0.000     1.00
```

Regression Equation

Response = 8.67 + 0.5753 Dose

QUADRATIC MODEL

Regression Analysis: Response versus Dose, Dose^2

```
Analysis of Variance

Source        DF   Adj SS   Adj MS    F-Value   P-Value
Regression     2   673.82   336.910   44.63     0.000
  Dose         1   205.97   205.970   27.29     0.000
  Dose^2       1    82.90    82.904   10.98     0.006
Error         12    90.58     7.548
  Lack-of-Fit  2    47.25    23.623    5.45     0.025
  Pure Error  10    43.33     4.333
Total         14   764.40
```

Model Summary

```
   S      R-sq   R-sq(adj)   R-sq(pred)
2.74741   88.15%   86.18%      80.70%
```

Coefficients

```
Term         Coef     SE Coef   T-Value   P-Value   VIF
Constant     4.48     1.66      2.71      0.019
Dose         1.506    0.288     5.22      0.000     19.67
Dose^2      -0.02699  0.00814  -3.31      0.006     19.67
```

Regression Equation

Response = 4.48 + 1.506 Dose - 0.02699 Dose^2

c. The quadratic model provides the better fit. The quadratic model has a much lower MS(Error), its R^2 value is 11% larger, the quadratic term has a p-value of 0.0062 which indicates that the term is significantly different from 0, however, the residual plot still has a distinct curvature as was found in the residual plot for the linear model.

12.11
 a. $y = 57.3 + 5.80$ Age $+ 3.32$ Weight

 b. $S = 2.45382$
    The estimated standard deviation can be found as S in the output or as $\sqrt{MS(Residual)}$

 c. β_2 can be interpreted as the amount of change in the average systolic blood pressure for a one pound increase in weight given that age is held constant. For a given age, an increase in one kilogram of weight is associated with an increase of 3.32 units in the average systolic blood pressure.

12.13
 a.
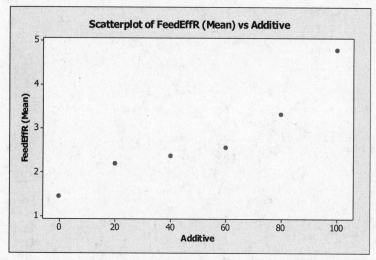

 b. It appears the cubic regression is the most appropriate as the data has a cubic shape.

 c. As expected, the cubic regression fits the best because the residuals in the lower order regression models exhibit a clear pattern. Also, the R^2 is significantly higher in the cubic model and all the coefficients of all three explanatory variables are significant at any reasonable α level.

 d. The data point 3.87 at additive level 100 does not fit the pattern of the data. It has a significantly lower value than the other FERs. Perhaps the data was input incorrectly or measured inaccurately.

12.15
Regression Analysis: Revenue versus Distance, Population

```
Analysis of Variance

Source        DF   Adj SS   Adj MS   F-Value   P-Value
Regression     2    35286    17643     14.65     0.000
  Distance     1     1494     1494      1.24     0.279
  Population   1    35243    35243     29.27     0.000
Error         19    22874     1204
Total         21    58160


Model Summary

      S    R-sq   R-sq(adj)   R-sq(pred)
34.6974   60.67%     56.53%       19.37%


Coefficients

Term        Coef   SE Coef   T-Value   P-Value    VIF
Constant   144.4      39.5      3.65     0.002
Distance   0.169     0.152      1.11     0.279   1.06
Population 1.094     0.202      5.41     0.000   1.06


Regression Equation

Revenue = 144.4 + 0.169 Distance + 1.094 Population

Fits and Diagnostics for Unusual Observations

Obs   Revenue    Fit   Resid   Std Resid
20      450.0  435.8    14.2        1.05   X
21      340.0  270.0    70.0        2.22   R
22      200.0  284.0   -84.0       -3.07   R

R  Large residual
X  Unusual X
```

a. Given the F=14.65 and a $p-value < 0.001$, we can reject the hypothesis of no overall predictive value.
b. Distance to the airport is not a significant predictor ($p = 0.279$)
c. $0.169 \pm 2.093(0.152) \Rightarrow 0.169 \pm 0.318 \Rightarrow (-0.149, 0.487)$
d. $z = \frac{1.094 - 0.5}{0.202} = 2.94 > 1.645$

Therefore, we have sufficient evidence the partial slope for population is significantly larger than 0.5.

12.17
a. The plot is shown below. There doesn't appear to be much of relationship among any of the variables.

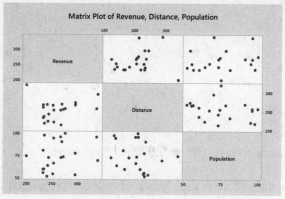

b. The regression output is below.

Regression Analysis: Revenue versus Distance, Population

```
Method

Rows unused   1


Analysis of Variance

Source        DF   Adj SS   Adj MS   F-Value   P-Value
Regression     2     2373     1187      0.99     0.391
  Distance     1     1060     1060      0.89     0.359
  Population   1     1843     1843      1.54     0.231
Error         18    21552     1197
Total         20    23925


Model Summary

      S    R-sq   R-sq(adj)   R-sq(pred)
34.6024   9.92%       0.00%        0.00%


Coefficients

Term          Coef   SE Coef   T-Value   P-Value    VIF
Constant     184.6      54.9      3.36     0.003
Distance     0.144     0.153      0.94     0.359   1.06
Population   0.617     0.497      1.24     0.231   1.06
```

128   Chapter 12: Multiple Regression and the General Linear Model

```
Regression Equation

Revenue = 184.6 + 0.144 Distance + 0.617 Population


Fits and Diagnostics for Unusual Observations

Obs  Revenue    Fit  Resid  Std Resid
 21    340.0  269.6   70.4       2.24  R
 22    200.0  279.9  -79.9      -2.96  R


R  Large residual
```

The model shows a very poor fit ($R^2 = 9.92\%$). Revenue is not related to distance from the hub.

c. Without plotting the data, there is the chance of mistakenly believing that population is related to revenue because the outlier is not discovered. The outlier is the driving factor in any factor being significant in this case.

12.19
Regression Analysis: Systolic versus Age, Weight

```
Analysis of Variance

Source            DF   Adj SS   Adj MS  F-Value  P-Value
Regression         4  1551.66  387.915    75.81    0.000
  Age              1    35.12   35.124     6.86    0.016
  Weight           1    20.40   20.400     3.99    0.060
  Age*Age          1     8.41    8.411     1.64    0.214
  Weight*Weight    1    13.83   13.834     2.70    0.116
Error             20   102.34    5.117
  Lack-of-Fit     16    85.84    5.365     1.30    0.439
  Pure Error       4    16.50    4.125
Total             24  1654.00
Model Summary

      S   R-sq  R-sq(adj)  R-sq(pred)
2.26210  93.81%     92.57%      89.21%


Coefficients

Term            Coef  SE Coef  T-Value  P-Value     VIF
Constant         9.6     20.8     0.46    0.649
Age            11.58     4.42     2.62    0.016  122.70
Weight          23.6     11.8     2.00    0.060  150.23
Age*Age       -0.649    0.506    -1.28    0.214  133.09
Weight*Weight  -2.84     1.72    -1.64    0.116  156.45
```

```
Regression Equation

Systolic = 9.6 + 11.58 Age + 23.6 Weight - 0.649 Age*Age - 2.84 Weight*Weight


Fits and Diagnostics for Unusual Observations

Obs  Systolic    Fit  Resid  Std Resid
  8    110.00 104.82   5.18       2.46  R
 16    103.00 104.45  -1.45      -1.33       X

R  Large residual
X  Unusual X
```

   a. Overall, the fit is quite good with an $R^2 = 93.81\%$. However, the model may have some uninformative explanatory variables due to large p-values for the two quadratic terms.

   b. As stated above, the quadratic terms do not appear useful to predict BP when the first order linear terms are included. The F-test of $H_0: \beta_3 = 0$ and $\beta_4 = 0$ vs. $H_a: \beta_3 \neq 0$ or $\beta_4 \neq 0$ where β_3 and β_4 are the coefficients of the quadratic terms is
$$F = \frac{(1551.66 - 1521.53)/2}{5.12} = 2.94 < 3.49 = F_{0.05,2,20}$$
Thus, H_0 is not rejected. There is not significant evidence that the quadratic terms are needed in the model.

   c. A first-order model appears to be more appropriate as the quadratic terms don't yield significant new information.

12.21

   a. 95% confidence intervals have the following form: $\beta_1 \pm t_{n-(k+1),0.025} SE(\beta_1)$
$t_{0.025,n-(k+1)} = t_{0.025,24-(3+1)} = t_{0.025,20} = 2.086$
Age: $11.576 \pm 2.086(4.419) \Rightarrow (2.358, 20.794)$
Weight: $23.62 \pm 2.086(11.83) \Rightarrow (-1.057, 48.297)$

   b. We are 95% confident the coefficient for age in the quadratic model is between 2.358 and 20.794. At the .05 level, we can see the coefficient is different from 0 (as 0 is not in the interval) and therefore, the age term is important.
We are 95% confident the coefficient for weight in the quadratic model is between -1.057 and 48.297. At the .05 level, we can see the coefficient is not significantly different from 0 (as 0 is slightly in the interval) and therefore, the weight term is marginal in importance.

12.23

   a. The variables weight, age, time, and pulse explain 58.2% of the variability in maximal oxygen uptake.

b. The regression output is shown below.

Regression Analysis: y versus x1, x2, x3, x4

```
Analysis of Variance

Source         DF   Adj SS    Adj MS    F-Value   P-Value
Regression      8   6.2213    0.777658    8.18    0.000
  x1            1   0.1754    0.175430    1.85    0.181
  x2            1   0.0095    0.009452    0.10    0.754
  x3            1   0.0008    0.000793    0.01    0.928
  x4            1   0.0031    0.003090    0.03    0.858
  x1*x1         1   0.0996    0.099618    1.05    0.312
  x2*x2         1   0.0133    0.013256    0.14    0.711
  x3*x3         1   0.0135    0.013501    0.14    0.708
  x4*x4         1   0.0026    0.002552    0.03    0.871
Error          45   4.2787    0.095083
Total          53  10.5000


Model Summary

      S     R-sq   R-sq(adj)   R-sq(pred)
0.308355   59.25%    52.01%      44.06%


Coefficients

Term             Coef    SE Coef   T-Value   P-Value      VIF
Constant         -4.3      14.3     -0.30     0.766
x1              0.0513    0.0378    1.36      0.181      171.68
x2              0.39      1.23      0.32      0.754     1178.51
x3              0.050     0.547     0.09      0.928      428.47
x4             -0.0049    0.0272   -0.18      0.858      120.05
x1*x1          -0.000127  0.000125 -1.02      0.312      170.65
x2*x2          -0.0108    0.0290   -0.37      0.711     1175.91
x3*x3          -0.0064    0.0170   -0.38      0.708      429.95
x4*x4          -0.000017  0.000102 -0.16      0.871      120.67


Regression Equation

y = -4.3 + 0.0513 x1 + 0.39 x2 + 0.050 x3 - 0.0049 x4 - 0.000127 x1*x1 -
 0.0108 x2*x2
     - 0.0064 x3*x3 - 0.000017 x4*x4

Fits and Diagnostics for Unusual Observations

Obs     y      Fit    Resid   Std Resid
 18   0.900   1.603  -0.703    -2.38   R
 29   1.300   1.124   0.176     0.92         X
 43   2.100   2.066   0.034     0.25         X
 49   2.500   1.888   0.612     2.05   R

R  Large residual
X  Unusual X
```

The R^2 only increased by 1.1%. The quadratic terms don't add much explanatory power to the model.

c. None of the partial slopes are significant in this model.

d. $Partial\ F = \frac{(6.2213-6.1062)/4}{4.2787/45} = 0.302$

This F is below any reasonable critical value meaning that the four quadratic terms taken as a whole are not useful predictors.

e. The first-order model is recommended because the second order model gives very little additional explanatory power and the quadratic predictors are not significant.

12.25
```
Coefficients

Term                              Coef     SE Coef   T-Value  P-Value    VIF
Constant                         1.659       0.127     13.09    0.000
Additive                       0.00267     0.00596      0.45    0.656   25.44
CopperLevel                   0.000116    0.000448      0.26    0.797    4.93
Additive*Additive             0.000258    0.000057      4.51    0.000   25.44
Additive*CopperLevel         -0.000001    0.000021     -0.04    0.966   52.69
Additive*Additive*CopperLevel 0.000000    0.000000      0.18    0.856   38.94
```

a. Only the quadratic term of the amount of the feed additive is significant.

b. Using the partial F-test

$$F = \frac{\frac{SSReg,Mod4 - SSReg,Mod3}{(k-g)}}{\frac{SSResidual,Mod4}{n-(k+1)}} = \frac{(63.2207-63.1959)/2}{5.2792/(60-6)} = 0.126$$

This is a very small F-value that is not significant. Therefore, the added predictors are not significant.

c. The conclusions in parts a and b are consistent. The interaction term between the additive amount and copper level as well as the quadratic term interaction are not significant.

12.27

a. $R^2 = 0.979566 \Rightarrow 97.96\%$ of the variation in Rating Score is accounted for in the model containing the three independent variables.

b. $F = \frac{0.979566/3}{(1-0.979566)/496} = 7925.76$ This value is slightly smaller than the value on the output due to rounding error.

c. The p-value associated with such a large F-value would be much less than 0.0001 and hence there is highly significant evidence that the model containing the three independent variables provides predictive value for the Rating Score.

12.29

a. $\hat{y} = 50.0195 + 6.64357x_1 + 7.3145x_2 - 1.23143x_1^2 - 0.7724x_1x_2 - 1.1755x_2^2$

b. $\hat{y} = 70.31 + 2.676x_1 - 0.8802x_2$

c. For the complete model: $R^2 = 86.24\%$
For the reduced model: $R^2 = 58.85\%$

d. In the complete model, we want to test
$H_0: \beta_3 = \beta_4 = \beta_5 = 0$ versus H_a: at least one of $\beta_3, \beta_4, \beta_5 \neq 0$.

The F-statistic has the form:
$$F = \frac{(SSReg, Complete - SSReg, Reduced)/(k - g)}{(SSResidual, Complete)/(n - (k + 1))} = \frac{(448.193 - 305.808)/(5 - 2)}{71.489/(20 - 6)} = 9.29$$
with $df = 3, 14 \Rightarrow p-value = \Pr(F_{3,14} \geq 9.29) = 0.0012 \Rightarrow$
Reject H_0. There is substantial evidence to conclude that at least one of $\beta_3, \beta_4, \beta_5 \neq 0$. Based on the F-test, omitting the second order terms from the model has substantially changed the fit of the model. Dropping one or more of these independent variables from the model will result in a decrease in the predictive value of the model.

12.31
a. Similar to the previous problem, we can use the output. $\widehat{BP} = 107.870$

b. For the 95% CI of the BP of an infant weighing 5 kg and 8 days old, we have from the output
$\Rightarrow (96.062, 119.679)$

12.33
a. Fit: 1.899 ; 95% PI: (1.284, 2.514)
b. Fit: 1.930 ; 95% PI: (1.292, 2.568)
c. Fit: 1.943 ; 95% PI: (1.279, 2.608)
d. Actually, the added complexity increased the width of the intervals because the MSE was increased (fewer df for error).

12.35
a.
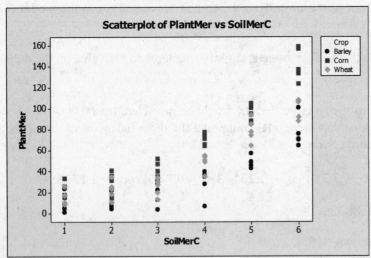
b. The relationship between soil mercury content and plant mercury content appears to be quadratic.
c. The relationship appears to be different at the first-order level, but it's very hard to distinguish if they have different quadratic terms. The relationship is similar with slightly different quadratic components (corn slopes up more quickly than barley).

Chapter 12: Multiple Regression and the General Linear Model

12.37
 a. $pc = -14.0218 + 18.084 mc$

 b. Wheat: $pc = -14.1773 + 17.8211 mc$
    Barley: $pc = -12.5283 + 12.8765 mc$
    Corn: $pc = -15.3600 + 23.5542 mc$

 c. These three equations have significantly different slopes but comparable intercepts as expected from the results of Exercise 12.39.

12.39
 a. $pc = 19.5727 - 7.1118 mc + 3.5994 MC^2$

 b. Wheat: $pc = 15.396 - 4.3589 mc + 3.1686 MC^2$
    Barley: $pc = 16.342 - 8.7759 mc + 3.0932 MC^2$
    Corn: $pc = 26.980 - 8.2008 mc + 4.5365 MC^2$

 c. These three equations appear to have different intercepts, linear slopes, and quadratic slopes, but the difference is not particularly significant due to the large standard deviations of the coefficients.

12.41
 a. For testing $H_0: \beta_1 = 0$ versus $H_a: \beta_1 \neq 0$, the p-value from the output is <0.0001. Thus, we can reject H_0 and conclude there is significant evidence that the probability of Tumor Development is related to the Amount of Additive.

 b. From the output, $\hat{p}(100) = 0.827$ with 95% C.I. (0.669, 0.919)

12.43 For this problem, the last y value (36) was deleted to make the matrices conformable.

 a. $X'X = \begin{bmatrix} 10.0 & 49.5 & 80.2 \\ 49.5 & 321.43 & 423.84 \\ 80.2 & 423.84 & 688.9 \end{bmatrix}$

 $(X'X)^{-1} = \begin{bmatrix} 0.1 & 0.0202 & 0.0125 \\ 0.0202 & 0.0031 & 0.0024 \\ 0.0125 & 0.0024 & 0.0015 \end{bmatrix}$

 $X'Y = \begin{bmatrix} 267.0 \\ 1269.5 \\ 2107.7 \end{bmatrix}$

 b. $\beta = \begin{bmatrix} \beta_0 \\ \beta_1 \\ \beta_2 \end{bmatrix} = (X'X)^{-1} X'Y = \begin{bmatrix} 78.627 \\ 14.316 \\ 9.384 \end{bmatrix}$

12.45 The Y matrix (vector) is the same for both models and is located in part b

a. X=

1	3	2.61000
1	4	2.67000
1	5	2.98000
1	6	3.98000
1	3	2.87000
1	4	3.41000
1	5	3.49000
1	6	4.03000
1	3	3.41000
1	4	2.81000
1	5	3.24000
1	6	3.75000
1	3	3.18000
1	4	3.13000
1	5	3.98000
1	6	4.55000
1	3	3.41000
1	4	3.35000
1	5	3.75000
1	6	3.83000
1	3	3.18000
1	4	3.52000
1	5	3.49000
1	6	3.81000
1	6	4.03000

b. X=

1	3	2.61	9	6.8121	7.8300
1	4	2.67	16	7.1289	10.6800
1	5	2.98	25	8.8804	14.9000
1	6	3.98	36	15.8404	23.8800
1	3	2.87	9	8.2369	8.6100
1	4	3.41	16	11.6281	13.6400
1	5	3.49	25	12.1801	17.4500
1	6	4.03	36	16.2409	24.1800
1	3	3.41	9	11.6281	10.2300
1	4	2.81	16	7.8961	11.2400
1	5	3.24	25	10.4976	16.2000
1	6	3.75	36	14.0625	22.5000
1	3	3.18	9	10.1124	9.5400
1	4	3.13	16	9.7969	12.5200
1	5	3.98	25	15.8404	19.9000
1	6	4.55	36	20.7025	27.3000
1	3	3.41	9	11.6281	10.2300
1	4	3.35	16	11.2225	13.4000
1	5	3.75	25	14.0625	18.7500
1	6	3.83	36	14.6689	22.9800
1	3	3.18	9	10.1124	9.5400
1	4	3.52	16	12.3904	14.0800
1	5	3.49	25	12.1801	17.4500
1	6	3.81	36	14.5161	22.8600
1	6	4.03	36	16.2409	24.1800

Y'=[80,90,96,102,81,96,99,110,88,90,100,102,86,93,101,103,86,91,100,105,84,91,95,104,107]

12.47

a. $F = \dfrac{0.894477/4}{(1-0.894477)/(43-5)} = 80.53$ with df = 4,38.

The $p-value = \Pr(F_{4,38} \geq 80.53) < 0.0001 \Rightarrow$ Reject $H_0: \beta_1 = \beta_2 = \beta_3 = \beta_4 = 0$ and conclude that at least one of the four independent variables has predictive value for Loan Volume.

b. Using $\alpha = 0.01$ none of the p-values for testing $H_0: \beta_i = 0$ versus $H_a: \beta_i \neq 0$; 0.0999, 0.0569, 0.5954, and 0.3648, respectively, are less than 0.01. Thus, none of the independent variables provide substantial predictive value given the remaining three variables in the model. That is, given a model with three variables included in the model, the fourth variable does not add much by including it also.

c. The contradiction is due to the severe collinearity that is present in the four independent variables. The F test demonstrates that as a group the four independent variables provide predictive value, but because the four independent variables are highly correlated, the information concerning their relationship with the dependent variable, Loan Volume, is highly overlapping. Thus, it is very difficult to determine which of the independent variables are useful in predicting Loan Volume.

12.49

Binary Logistic Regression: Yes, Trials versus Dust, Race, ...

Link Function: Logit

Response Information

```
Variable  Value       Count
Yes       Event         165
          Non-event    5254
Trials    Total        5419
```

* NOTE * 65 cases were used
* NOTE * 7 cases contained missing values

Logistic Regression Table

```
                                                Odds      95% CI
Predictor      Coef    SE Coef        Z      P  Ratio  Lower  Upper
Constant   -1.94520   0.233415    -8.33  0.000
Dust
  2        -2.57991   0.292058    -8.83  0.000   0.08   0.04   0.13
  3        -2.73063   0.215338   -12.68  0.000   0.07   0.04   0.10
Race
  2         0.116349  0.207194     0.56  0.574   1.12   0.75   1.69
Sex
  2         0.123930  0.228754     0.54  0.588   1.13   0.72   1.77
Smoking
  2        -0.641347  0.194422    -3.30  0.001   0.53   0.36   0.77
Employ
  2         0.564111  0.261658     2.16  0.031   1.76   1.05   2.94
  3         0.753138  0.216141     3.48  0.000   2.12   1.39   3.24
```

Log-Likelihood = -598.968
Test that all slopes are zero: G = 279.256, DF = 7, P-Value = 0.000

```
Goodness-of-Fit Tests

Method              Chi-Square    DF      P
Pearson               37.9336     57    0.976
Deviance.             43.2710     57    0.910
Hosmer-Lemeshow        7.0633      7    0.422


Table of Observed and Expected Frequencies:
(See Hosmer-Lemeshow Test for the Pearson Chi-Square Statistic)

Value         Event            Non-event
Group   Observed  Expected  Observed  Expected   Total
  1        7        4.4       761      763.6      768
  2        7        4.6       559      561.4      566
  3        8        5.6       537      539.4      545
  4        5        7.3       651      648.7      656
  5        7        8.4       675      673.6      682
  6       18       18.2      1006     1005.8     1024
  7       14       19.0       592      587.0      606
  8       99       97.0       471      473.0      570
  9        0        0.5         2        1.5        2


Measures of Association:
(Between the Response Variable and Predicted Probabilities)

Pairs          Number   Percent   Summary Measures
Concordant     660072    76.1     Somers' D              0.58
Discordant     153811    17.7     Goodman-Kruskal Gamma  0.62
Ties            53027     6.1     Kendall's Tau-a        0.03
Total          866910   100.0
```

a. From the output, based on the p-values of the independent variables in the model, the most likely to be associated with brown lung disease is Dust, followed by Employ, Smoking, Sex, and finally Race.

b. SAS Output for a logistic model with all 2-way interactions is shown here:

Model Convergence Status
Convergence criterion (GCONV=1E-8) satisfied.

Model Fit Statistics

Criterion	Intercept Only	Intercept and Covariates
AIC	1479.192	1224.332
SC	1481.827	1295.470
-2 Log L	1477.192	1170.332

Testing Global Null Hypothesis: BETA=0

Test	Chi-Square	DF	Pr > ChiSq
Likelihood Ratio	306.8601	26	<.0001
Score	565.6458	26	<.0001
Wald	303.2090	26	<.0001

Type 3 Analysis of Effects

Effect	DF	Wald Chi-Square	Pr > ChiSq
Dust	2	27.6884	<.0001
Race	1	0.9357	0.3334
Sex	1	0.2307	0.6310
Smoking	1	1.7264	0.1889
Employ	2	0.4022	0.8178
Dust*Race	2	1.7920	0.4082
Dust*Sex	2	4.2512	0.1194
Dust*Smoking	2	3.7978	0.1497
Dust*Employ	4	4.2757	0.3700
Race*Sex	1	0.4507	0.5020
Race*Smoking	1	1.3366	0.2476
Race*Employ	2	5.9673	0.0506

Chapter 12: Multiple Regression and the General Linear Model

Type 3 Analysis of Effects			
Effect	DF	Wald Chi-Square	Pr > ChiSq
Sex*Smoking	1	0.2220	0.6375
Sex*Employ	2	2.3556	0.3080
Smoking*Employ	2	0.2013	0.9043

Analysis of Maximum Likelihood Estimates						
Parameter			DF	Estimate	Standard Error	Wald Chi-Square	Pr > ChiSq
Intercept			1	-4.0986	0.2582	251.9282	<.0001
Dust	1		1	1.5075	0.3057	24.3114	<.0001
Dust	2		1	-0.5778	0.2914	3.9316	0.0474
Race	1		1	0.1971	0.2037	0.9357	0.3334
Sex	1		1	0.0992	0.2065	0.2307	0.6310
Smoking	1		1	0.1721	0.1310	1.7264	0.1889
Employ	1		1	-0.1327	0.2348	0.3192	0.5721
Employ	2		1	0.0330	0.2596	0.0162	0.8987
Dust*Race	1	1	1	-0.2930	0.2914	1.0110	0.3147
Dust*Race	2	1	1	0.0477	0.3213	0.0221	0.8819
Dust*Sex	1	1	1	0.5376	0.2738	3.8556	0.0496
Dust*Sex	2	1	1	-0.4572	0.2458	3.4607	0.0628
Dust*Smoking	1	1	1	0.3607	0.1906	3.5793	0.0585
Dust*Smoking	2	1	1	-0.3360	0.2045	2.6989	0.1004
Dust*Employ	1	1	1	-0.4355	0.3921	1.2339	0.2667
Dust*Employ	1	2	1	0.4008	0.3676	1.1887	0.2756
Dust*Employ	2	1	1	-0.0896	0.4203	0.0455	0.8312
Dust*Employ	2	2	1	0.1732	0.4000	0.1875	0.6650
Race*Sex	1	1	1	-0.1386	0.2065	0.4507	0.5020

Chapter 12: Multiple Regression and the General Linear Model

Analysis of Maximum Likelihood Estimates						
Parameter			DF	Estimate	Standard Error	Wald Chi-Square	Pr > ChiSq
Race*Smoking	1	1	1	0.1591	0.1376	1.3366	0.2476
Race*Employ	1	1	1	-0.5601	0.2299	5.9365	0.0148
Race*Employ	1	2	1	0.3683	0.2094	3.0935	0.0786
Sex*Smoking	1	1	1	-0.0674	0.1430	0.2220	0.6375
Sex*Employ	1	1	1	-0.3136	0.2706	1.3431	0.2465
Sex*Employ	1	2	1	0.0251	0.3006	0.0069	0.9336
Smoking*Employ	1	1	1	0.0248	0.1830	0.0183	0.8923
Smoking*Employ	1	2	1	-0.0831	0.2000	0.1728	0.6776

Association of Predicted Probabilities and Observed Responses			
Percent Concordant	60.6	Somers' D	0.271
Percent Discordant	33.5	Gamma	0.288
Percent Tied	5.9	Tau-a	0.127
Pairs	2470	c	0.635

Some of the interaction terms are significant so we will keep the interactions in the model.

c. The model is fit with all main effects and interactions

12.51

a. $\hat{y} = 0.8727 + 2.548\ SIZE + 0.220\ PARKING + 0.589\ INCOME$
   SE: (1.946)   (1.201)   (0.155)   (0.178)

b. The interpretation of the coefficients is given here:

Coefficient	Interpretation
$\hat{\beta}_0$ = y-intercept	The estimated average daily sales for the population of stores having 0 Size, 0 Parking, and 0 Income.
$\hat{\beta}_1 = \hat{\beta}_{SIZE}$	The estimated change in average Daily Sales per unit change in SIZE, for fixed values of PARKING and INCOME.
$\hat{\beta}_1 = \hat{\beta}_{PARKING}$	The estimated change in average Daily Sales per unit change in PARKING, for fixed values of SIZE and INCOME.
$\hat{\beta}_1 = \hat{\beta}_{INCOME}$	The estimated change in average Daily Sales per unit change in, INCOME for fixed values of SIZE and PARKING.

140   Chapter 12: Multiple Regression and the General Linear Model

   c. $R^2 = 0.7912$ and $s_e = 0.7724$

   d. Only the pairwise correlations between the independent variables are given on the output. A better indicator of collinearity is the value of VIF or the R^2 values from predicting each independent variable from the remaining independent variables. Examining the correlations does not reveal any large values. Only SIZE and PARKING with a correlation of 0.6565 appear to be near a value which would be a concern relative to collinearity.

12.53
   a. $\hat{y} = 102.708 - 0.833\ PROTEIN - 4.000\ ANTIBIO - 1.375\ SUPPLEM$
   b. $s_e = 1.70956$
   c. $R^2 = 90.07\%$
   d. There is no collinearity problem in the data set. The correlations between the pairs of independent variables are 0 for each pair and the VIF values are all equal to 1.0. This total lack of collinearity is due to the fact that the independent variables are perfectly balanced. Each combination of PROTEIN and ANTIBIO values appears exactly three times in the data set. Each combination of PROTEIN and SUPPLEM occurs twice, etc.

12.55
   a. $\hat{y} = 89.8333 - 0.83333\ PROTEIN$
   b. $R^2 = 0.5057$
   c. In the complete model, we want to test
$$H_0: \beta_2 = \beta_3 = 0 \text{ versus } H_a: \text{at least one of } \beta_2, \beta_3 \neq 0.$$
The F-statistic has the form:
$$F = \frac{(371.083 - 208.333)/(3 - 1)}{40.9166/(18 - 4)} = 27.84$$
With $df = 2,14 \Rightarrow p-value = \Pr(F_{2,14} \geq 27.84) < 0.0001 \Rightarrow$ Reject H_0
Based on the F-test, omitting ANTIBIO and/or SUPPLEM from the model would substantially change the fit of the model. Dropping ANTIBIO and/or SUPPLEM from the model may result in a large decrease in the predictive value of the model.

12.57
   a. $R^2 = 0.3844 = 38.44\%$
   b. The Minitab output is given here:

Regression Analysis: SALARY versus NUMEXPL

```
The regression equation is
SALARY = 29.1 + 0.00326 NUMEXPL

Predictor      Coef    SE Coef       T        P
Constant    29.0841     0.2348  123.85    0.000
NUMEXPL    0.003259   0.002098    1.55    0.125

S = 1.23171    R-Sq = 3.6%    R-Sq(adj) = 2.1%

Analysis of Variance

Source             DF        SS       MS       F       P
Regression          1     3.662    3.662    2.41   0.125
Residual Error     65    98.612    1.517
Total              66   102.274
```

```
Unusual Observations

Obs  NUMEXPL  SALARY     Fit  SE Fit  Residual  St Resid
 12      389  28.900  30.352   0.653    -1.452     -1.39 X
 46      130  25.700  29.508   0.177    -3.808     -3.12R
 56      371  32.400  30.293   0.617     2.107      1.98 X
 64      279  29.400  29.993   0.432    -0.593     -0.51 X
```

R denotes an observation with a large standardized residual.
X denotes an observation whose X value gives it large leverage.

R^2 has decreased dramatically to 0.0358=3.58%

c. In the complete model, we want to test
$$H_0: \beta_2 = \beta_3 = 0 \text{ versus } H_a: \text{ at least one of } \beta_2, \beta_3 \neq 0.$$
The F-statistic has the form:
$$F = \frac{(39.31706 - 3.66167)/(3-1)}{62.95698/(67-4)} = 17.84$$
With $df = 2, 63 \Rightarrow p - value = \Pr(F_{2,63} \geq 17.84) < 0.0001 \Rightarrow$ Reject H_0

There is substantial evidence to conclude that at least one of $\beta_2, \beta_3 \neq 0$. Based on the F-test, omitting MARGIN and/or IPCOST from the model would substantially change the fit of the model. Dropping MARGIN and/or IPCOST from the model may result in a large decrease in the predictive value of the model.

12.59 The Minitab Output is given here:

Correlations: OVERCOST, NUMEMPL, SIZE, PERSCOST

```
          OVERCOST   NUMEMPL      SIZE
NUMEMPL      0.426
             0.000

SIZE         0.379     0.342
             0.000     0.001

PERSCOST    -0.320    -0.027    -0.140
             0.003     0.802     0.199

Cell Contents: Pearson correlation
               P-Value
```

MODEL WITH OUTLIER
Regression Analysis: OVERCOST versus NUMEMPL, SIZE, PERSCOST

```
The regression equation is
OVERCOST = 78.4 + 0.00264 NUMEMPL + 1.58 SIZE - 0.245 PERSCOST

Predictor         Coef      SE Coef         T      P
Constant        78.351        5.341     14.67  0.000
NUMEMPL       0.0026371    0.0007484      3.52  0.001
SIZE            1.5763       0.6918      2.28  0.025
PERSCOST      -0.24481      0.08072     -3.03  0.003


S = 6.03159    R-Sq = 31.9%    R-Sq(adj) = 29.4%

Analysis of Variance

Source           DF        SS       MS       F      P
Regression        3   1398.47   466.16   12.81  0.000
Residual Error   82   2983.16    36.38
Total            85   4381.63

Source       DF    Seq SS
NUMEMPL       1    793.43
SIZE          1    270.45
PERSCOST      1    334.59

Unusual Observations

Obs  NUMEMPL  OVERCOST     Fit  SE Fit  Residual  St Resid
 17      126    56.400  68.601   0.934   -12.201    -2.05R
 26      194    78.100  64.535   1.131    13.565     2.29R
 32     7216    76.400  86.992   4.722   -10.592    -2.82RX
 49     1866    85.600  77.209   2.296     8.391     1.50 X
 68      214    76.700  63.706   1.730    12.994     2.25R
 70     3766    80.900  79.847   2.314     1.053     0.19 X
 75      769    83.600  70.920   0.900    12.680     2.13R
```

R denotes an observation with a large standardized residual.
X denotes an observation whose X value gives it large leverage.

MODEL WITHOUT OUTLIER

Regression Analysis: OVERCOST versus NUMEMPL, SIZE, PERSCOST

```
The regression equation is
OVERCOST = 75.0 + 0.00516 NUMEMPL + 1.36 SIZE - 0.205 PERSCOST

Predictor       Coef     SE Coef       T      P
Constant      75.048       5.227   14.36  0.000
NUMEMPL      0.005160    0.001115   4.63  0.000
SIZE          1.3632      0.6653    2.05  0.044
PERSCOST    -0.20498     0.07835   -2.62  0.011

S = 5.76646    R-Sq = 37.7%    R-Sq(adj) = 35.4%

Analysis of Variance

Source          DF        SS       MS       F      P
Regression       3   1630.78   543.59   16.35  0.000
```

```
Residual Error  81  2693.41  33.25
Total           84  4324.19

Source     DF   Seq SS
NUMEMPL     1  1215.98
SIZE        1   187.18
PERSCOST    1   227.62
```

Unusual Observations

```
Obs  NUMEMPL  OVERCOST    Fit   SE Fit  Residual  St Resid
 26      194    78.100  64.090   1.091    14.010   2.47R
 36      157    80.100  67.258   1.063    12.842   2.27R
 48     1866    85.600  79.683   2.350     5.917   1.12 X
 67      214    76.700  63.531   1.655    13.169   2.38R
 69     3766    80.900  87.556   3.423    -6.656  -1.43 X
 74      769    83.600  71.193   0.865    12.407   2.18R
```

R denotes an observation with a large standardized residual.
X denotes an observation whose X value gives it large leverage.

a. None of the correlations among independent variables are particularly large. The largest value is between NUMEMPL and SIZE but this value is only 0.34085. Thus, unless there is a somewhat complex relationship between several of the independent variables, there would not appear to be any severe collinearity problems.

b. The scatterplots of the data are given here:

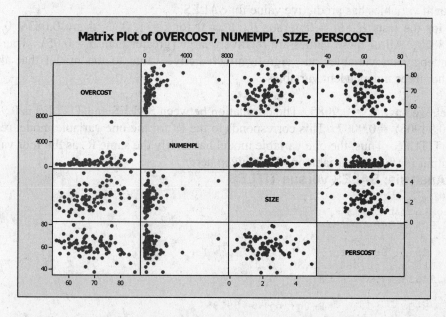

A possible outlier is observed in the plot of OVERHEAD versus NUMEMPL. One data value is plotted far to the right of the remaining data values, so it has a high leverage value. The data would be far below any fitted line.

c. $\hat{y} = 78.35149 + 0.00264\ NUMEMPL + 1.5763\ SIZE - 0.24481\ PERSCOST$

d. With the outlier removed from the dataset, we obtain
$$\hat{y} = 75.04849 + 0.00516\ NUMEMPL + 1.3632\ SIZE - 0.20498\ PERSCOST$$
The slope for NUMEMPL nearly doubled in size after removing the outlier but the other two slopes did not change a great amount.

12.61 The Minitab Output is given below:
```
Predicted Values for New Observations

New
Obs    Fit   SE Fit     95% CI          95% PI
  1  69.763  0.787  (68.197, 71.328)  (58.183, 81.342)

Values of Predictors for New Observations

New
Obs  NUMEMPL  SIZE   PERSCOST
  1    500    2.50     55.0
```

From the output, we have that when NUMEMPL=500, SIZE=2.5, and PERSCOST=55, $\hat{y}$ = 69.763% and a 95% P.I. for y is (58.183, 81.342). The value of 88.9% falls outside the P.I. and hence would appear to be somewhat unreasonable in this situation.

12.63
a. To test $H_0: \beta_1 = \beta_2 = \beta_3 = \beta_4 = 0$ versus H_a: at least one of $\beta_i \neq 0$, we can use the F-test from the output: F=116.68 with p-value <0.0001. Yes we can reject H_0 and conclude that at least one of the independent variables has predictive value for SALES.

b. The p-values for the tests $H_0: \beta_i = 0$ versus $H_a: \beta_i \neq 0$ for i = 1, 2, 3, 4 are 0.0836, 0.1024, 0.1799, and 0.8827. All of these values are relatively large (greater than $\alpha = 0.05$). Therefore, none of the independent variables add significant predictive value to the model that already contains the other three independent variables.

12.65 From the model, we have $R^2 = 0.9085$. The correlation between SALES and TITLES is 0.94905. It has square $(0.94905)^2 = 0.9007$. This corresponds to the R^2 for the one variable model relating SALES just to TITLES. Thus, the one variable model has nearly the same R^2 as the four variable model. The output for the one variable model is given here:

Regression Analysis: SALES versus TITLES

```
The regression equation is
SALES = - 10.9 + 0.846 TITLES

Predictor     Coef   SE Coef      T      P
Constant   -10.888     5.880  -1.85  0.070
TITLES     0.84633   0.03974  21.30  0.000

S = 6.23502   R-Sq = 90.1%   R-Sq(adj) = 89.9%

Analysis of Variance

Source          DF      SS     MS       F      P
Regression       1   17632  17632  453.55  0.000
Residual Error  50    1944     39
Total           51   19576
```

```
Unusual Observations

Obs  TITLES    SALES      Fit   SE Fit  Residual  St Resid
 39     167  115.900  130.449    1.192   -14.549    -2.38R
 43     175  157.100  137.220    1.430    19.880     3.28R
 44     178  152.700  139.759    1.526    12.941     2.14R
```

R denotes an observation with a large standardized residual.

In the complete model, we want to test
$$H_0: \beta_2 = \beta_3 = \beta_4 = 0 \text{ versus } H_a: \text{at least one of } \beta_2, \beta_3, \beta_4 \neq 0.$$
The F-statistic has the form:
$$F = \frac{(17785 - 17632)/(4-1)}{1790.93411/(52-5)} = 1.34$$

With $df = 3, 47 \Rightarrow p-value = \Pr(F_{3,47} \geq 1.34) = 0.2727 \Rightarrow$ Fail to Reject H_0

There is not substantial evidence to conclude that at least one of $\beta_2, \beta_3, \beta_4 \neq 0$. Based on the F-test, omitting FOOTAGE, IBMBASE, and APLBASE from the model would not substantially change the fit of the model. Dropping FOOTAGE, IBMBASE, and APLBASE from the model will not result in a large decrease in the predictive value of the model. These variables are so highly correlated with TITLES that they add essentially no additional predictive value once SALES is modeled by TITLES. Also, there is the possibility that autocorrelation is confusing the issue. To truly relate SALES to the four independent variables, we would need data from 52 different computer stores during a fixed time period.

12.67

Model Fit Statistics		
Criterion	Intercept Only	Intercept and Covariates
AIC	8326.902	8327.647
SC	8328.962	8358.554
-2 Log L	8324.902	8297.647

Testing Global Null Hypothesis: BETA=0			
Test	Chi-Square	DF	Pr > ChiSq
Likelihood Ratio	27.2544	14	0.0178
Score	27.7882	14	0.0152
Wald	27.5123	14	0.0165

Type 3 Analysis of Effects			
Effect	DF	Wald Chi-Square	Pr > ChiSq
SocClass	4	8.4621	0.0760
SmokeLevel	2	1.8927	0.3882
SocClass*SmokeLevel	8	2.9953	0.9346

Analysis of Maximum Likelihood Estimates						
Parameter			DF	Estimate	Standard Error	Wald Chi-Square	Pr > ChiSq
Intercept			1	-2.2938	0.0909	636.6042	<.0001
SocClass	1		1	-0.0529	0.2967	0.0318	0.8586
SocClass	2		1	-0.3222	0.1874	2.9581	0.0854
SocClass	3		1	-0.0142	0.1012	0.0198	0.8881
SocClass	4		1	0.0760	0.1169	0.4229	0.5155
SmokeLevel	1		1	0.1243	0.0962	1.6681	0.1965
SmokeLevel	2		1	0.0215	0.1036	0.0431	0.8355
SocClass*SmokeLevel	1	1	1	0.2062	0.3060	0.4539	0.5005
SocClass*SmokeLevel	1	2	1	0.0739	0.3355	0.0485	0.8256
SocClass*SmokeLevel	2	1	1	-0.0349	0.1967	0.0315	0.8592
SocClass*SmokeLevel	2	2	1	0.0662	0.2116	0.0979	0.7544
SocClass*SmokeLevel	3	1	1	-0.0822	0.1083	0.5753	0.4482
SocClass*SmokeLevel	3	2	1	-0.0796	0.1163	0.4681	0.4939
SocClass*SmokeLevel	4	1	1	0.0711	0.1290	0.3041	0.5813
SocClass*SmokeLevel	4	2	1	-0.0660	0.1373	0.2308	0.6309

a. From the above Minitab output, the equation predicting Hypertension (assuming we maintain all interactions) is

$$\ln\left(\frac{p}{1-p}\right) = -2.29 - 0.051 I_{Soc=1} - 0.32 I_{Soc=2} - 0.014 I_{Soc=3} + 0.076 I_{Soc=4} + 0.12 I_{Smoke=1} + 0.02 I_{Smoke=2}$$
$$+ 0.21 I_{Soc=1, Smoke=1} + 0.07 I_{Soc=1, Smoke=2} - 0.03 I_{Soc=2, Smoke=1} + 0.07 I_{Soc=2, Smoke=2}$$
$$- 0.08 I_{Soc=3, Smoke=1} - 0.08 I_{Soc=3, Smoke=2} + 0.13 I_{Soc=4, Smoke=1} - 0.07 I_{Soc=4, Smoke=2}$$

b. $\ln\left(\frac{p}{1-p}\right) = -2.29 - 0.014 = -2.304 \Rightarrow p = 0.091$

c. SAS output gives 95% P.I. as (0.0699, 0.1197)

12.69

Model Fit Statistics		
Criterion	Intercept Only	Intercept and Covariates
AIC	5291.095	5278.211
SC	5293.155	5309.117
-2 Log L	5289.095	5248.211

Testing Global Null Hypothesis: BETA=0			
Test	Chi-Square	DF	Pr > ChiSq
Likelihood Ratio	40.8842	14	0.0002
Score	37.7608	14	0.0006
Wald	34.1895	14	0.0019

Type 3 Analysis of Effects			
Effect	DF	Wald Chi-Square	Pr > ChiSq
SocClass	4	4.3129	0.3653
SmokeLevel	2	5.8284	0.0542
SocClass*SmokeLevel	8	4.8635	0.7721

Analysis of Maximum Likelihood Estimates							
Parameter			DF	Estimate	Standard Error	Wald Chi-Square	Pr > ChiSq
Intercept			1	-3.7932	16.9853	0.0499	0.8233
SocClass	1		1	0.9933	16.9878	0.0034	0.9534
SocClass	2		1	-3.3795	67.9401	0.0025	0.9603
SocClass	3		1	0.6532	16.9854	0.0015	0.9693
SocClass	4		1	0.7046	16.9856	0.0017	0.9669
SmokeLevel	1		1	0.9914	16.9853	0.0034	0.9535
SmokeLevel	2		1	0.6471	16.9854	0.0015	0.9696
SocClass*SmokeLevel	1	1	1	-0.8228	16.9881	0.0023	0.9614
SocClass*SmokeLevel	1	2	1	-0.8429	16.9892	0.0025	0.9604
SocClass*SmokeLevel	2	1	1	3.1041	67.9402	0.0021	0.9636
SocClass*SmokeLevel	2	2	1	3.1168	67.9403	0.0021	0.9634
SocClass*SmokeLevel	3	1	1	-0.6188	16.9855	0.0013	0.9709
SocClass*SmokeLevel	3	2	1	-0.7037	16.9856	0.0017	0.9670
SocClass*SmokeLevel	4	1	1	-0.5851	16.9858	0.0012	0.9725
SocClass*SmokeLevel	4	2	1	-0.7173	16.9860	0.0018	0.9663

a. From the above Minitab output, the equation predicting both hypertension and proteinuria is found by utilizing the same methods in 12.70 with the new output given above. The model used will include all interactions and main effects.

b. Using SAS, the probability of both hypertension and proteinuria in a pregnant woman of social class 2 smoking between 1 and 19 cigarettes per day is 0.032.

c. SAS output gives 95% P.I. as (0.018, 0.054)

12.71

a. Regression results shown below.

Regression Analysis: Catch versus Homes, Lakesize, Structure, Access

```
Method

Categorical predictor coding   (1, 0)
```

Analysis of Variance

Source	DF	Adj SS	Adj MS	F-Value	P-Value
Regression	7	21.4060	3.05800	14.90	0.000
Homes	1	0.0767	0.07675	0.37	0.552
Lakesize	1	0.0449	0.04489	0.22	0.648
Structure	1	7.5592	7.55925	36.84	0.000
Access	1	0.0160	0.01603	0.08	0.785
Homes*Access	1	0.1717	0.17165	0.84	0.378
Lakesize*Access	1	0.0034	0.00341	0.02	0.900
Structure*Access	1	0.0553	0.05529	0.27	0.613
Error	12	2.4620	0.20517		
Total	19	23.8680			

Model Summary

S	R-sq	R-sq(adj)	R-sq(pred)
0.452954	89.68%	83.67%	69.63%

Coefficients

Term	Coef	SE Coef	T-Value	P-Value	VIF
Constant	-1.44	2.06	-0.70	0.497	
Homes	0.0121	0.0197	0.61	0.552	11.18
Lakesize	0.430	0.920	0.47	0.648	44.72
Structure	0.04782	0.00788	6.07	0.000	2.29
Access					
Yes	-0.68	2.42	-0.28	0.785	136.85
Homes*Access					
Yes	0.0237	0.0260	0.91	0.378	72.63
Lakesize*Access					
Yes	0.126	0.975	0.13	0.900	69.53
Structure*Access					
Yes	-0.0060	0.0115	-0.52	0.613	10.56

b. The formulas are shown here.

```
Regression Equation

Access
No      Catch = -1.44 + 0.0121 Homes + 0.430 Lakesize + 0.04782 Structure

Yes     Catch = -2.12 + 0.0358 Homes + 0.556 Lakesize + 0.04183 Structure
```

c. There is not a significant difference between the partial slopes for residency, size, and structure for lakes with and without access. The p-values are all large.

12.73
a. The regression output is shown below.

Regression Analysis: Catch versus Homes, Lakesize, Structure, Access

Method

Categorical predictor coding (1, 0)

Analysis of Variance

Source	DF	Adj SS	Adj MS	F-Value	P-Value
Regression	16	22.8194	1.42621	4.08	0.136
Homes	1	0.0025	0.00249	0.01	0.938
Lakesize	1	0.0002	0.00015	0.00	0.985
Structure	1	0.0012	0.00117	0.00	0.957
Access	1	0.0003	0.00026	0.00	0.980
Homes*Homes	1	0.0018	0.00179	0.01	0.947
Lakesize*Lakesize	1	0.0014	0.00142	0.00	0.953
Structure*Structure	1	0.0046	0.00463	0.01	0.916
Homes*Lakesize	1	0.0022	0.00224	0.01	0.941
Homes*Structure	1	0.0019	0.00189	0.01	0.946
Lakesize*Structure	1	0.0157	0.01570	0.04	0.846
Homes*Access	1	0.0000	0.00000	0.00	0.999
Lakesize*Access	1	0.0244	0.02441	0.07	0.809
Structure*Access	1	0.0290	0.02897	0.08	0.792
Homes*Homes*Access	1	0.0001	0.00008	0.00	0.989
Lakesize*Lakesize*Access	1	0.0080	0.00796	0.02	0.890
Structure*Structure*Access	1	0.0245	0.02450	0.07	0.808
Error	3	1.0486	0.34955		
Total	19	23.8680			

Model Summary

S	R-sq	R-sq(adj)	R-sq(pred)
0.591226	95.61%	72.17%	0.00%

Coefficients

Term	Coef	SE Coef	T-Value	P-Value	VIF
Constant	-7.6	61.3	-0.12	0.909	
Homes	0.17	2.00	0.08	0.938	67071.81
Lakesize	0.3	15.7	0.02	0.985	7632.82
Structure	0.040	0.693	0.06	0.957	10407.54
Access					
Yes	2.1	76.7	0.03	0.980	80859.81
Homes*Homes	-0.0010	0.0143	-0.07	0.947	63397.19
Lakesize*Lakesize	0.56	8.83	0.06	0.953	20587.59
Structure*Structure	-0.00029	0.00252	-0.12	0.916	1654.83
Homes*Lakesize	-0.011	0.136	-0.08	0.941	926.73
Homes*Structure	0.00037	0.00503	0.07	0.946	5136.93
Lakesize*Structure	0.0188	0.0886	0.21	0.846	670.59
Homes*Access					
Yes	-0.00	2.35	-0.00	0.999	350761.45
Lakesize*Access					
Yes	3.1	11.6	0.26	0.809	5826.76
Structure*Access					
Yes	-0.160	0.555	-0.29	0.792	14323.50

```
Homes*Homes*Access
  Yes                          0.0002    0.0157    0.01    0.989   98996.50
Lakesize*Lakesize*Access
  Yes                          -1.31     8.67      -0.15   0.890   21023.17
Structure*Structure*Access
  Yes                          0.00139   0.00524   0.26    0.808   7175.51
```

b. The fitted lines are shown below.
```
Access
No
Catch = -7.6 + 0.17 Homes + 0.3 Lakesize + 0.040 Structure - 0.0010 Homes*Homes
+ 0.56 Lakesize*Lakesize - 0.00029 Structure*Structure - 0.011 Homes*Lakesize
+ 0.00037 Homes*Structure + 0.0188 Lakesize*Structure

Yes
Catch = -5.6 + 0.167 Homes + 3.40 Lakesize - 0.119 Structure -
 0.00079 Homes*Homes - 0.75 Lakesize*Lakesize + 0.00110 Structure*Structure -
 0.011 Homes*Lakesize + 0.00037 Homes*Structure + 0.0188 Lakesize*Structure
```

c. All second order terms are highly insignificant so there is no improvement.

12.75 Fitting a complete second-order model would require an additional 6 parameters (3 interactions with and without the indicator variable) and there are only 3 remaining degrees of freedom with which to estimate parameters.

Chapter 13

Further Regression Topics

13.1 Individual Solutions

13.3
  a.
   Correlations: CRIME, AGE, COLLEGE, INCOME, GENDER
   ```
              CRIME      AGE  COLLEGE   INCOME
   AGE        0.243
              0.108

   COLLEGE    0.042    0.902
              0.784    0.000

   INCOME     0.815   -0.138   -0.342
              0.000    0.368    0.021

   GENDER     0.504   -0.110   -0.119    0.408
              0.000    0.474    0.438    0.005

   Cell Contents: Pearson correlation
                  P-Value
   ```

   There is a collinearity issue between student age and number of years in college (as expected) and also between CRIME and INCOME. Correlations of 0.902 and 0.815 are large and could lead to inflated standard errors.

  b. The Minitab output is given below:
   Best Subsets Regression: CRIME versus AGE, COLLEGE, INCOME, GENDER
   ```
   Response is CRIME
                                        C
                                        O   I  G
                                        L   N  E
                                        L   C  N
                                    A   E   O  D
                          Mallows   G   G   M  E
   Vars  R-Sq  R-Sq(adj)    Cp      S   E   E  E  R
     1   66.5    65.7      41.1   3.5023         X
     1   25.4    23.6     141.7   5.2239            X
     2   79.3    78.3      11.7   2.7843  X    X
     2   78.1    77.1      14.5   2.8609  X X
     3   83.6    82.5       3.0   2.5041  X       X  X
     3   81.3    80.0       8.7   2.6756  X   X   X
     4   83.7    82.0       5.0   2.5345  X X X X
   ```

   Using adjusted R^2, the optimal model includes age of student, income of parents, and gender.
  c. Individual Solutions

13.5
  a. Number of customers and square feet appear to be highly correlated. This is to be expected as more customers can visit a larger store.
  b. There is one point of high leverage in square feet (low sales in a large store) and one point of high leverage in number of customers (low sales with many customers). There is a possibility that this is the same point in both plots (likely as the value of Sales appears to be the same in both situations).
  c. Perhaps a correlation matrix and the calculation of the VIF would be in order to examine collinearity.

13.7 The variable x_4^2 was highly correlated with the other predictors. So much so, it was removed from the best subsets procedure. The Minitab output is given here:

Best Subsets Regression: y versus x1, x2, ...

```
Response is y
                                                  x x x
                                                  1 2 3 x x x x x x
                                                  s s s 1 1 1 2 2 3
                            Mallows             x x x x q q q x x x x x x
     Vars   R-Sq  R-Sq(adj)    Cp        S    1 2 3 4 r r r 2 3 4 3 4 4
       1    64.3     63.1    51.3    0.45461                       X
       1    58.4     57.0    64.5    0.49099  X
       2    75.5     73.8    28.4    0.38284                   X   X
       2    75.0     73.3    29.5    0.38674  X                    X
       3    81.5     79.5    17.1    0.33876          X            X   X
       3    79.7     77.5    21.2    0.35510  X       X            X
       4    84.8     82.6    11.7    0.31256          X            X X X
       4    83.0     80.5    15.7    0.33027  X       X            X   X
       5    88.1     85.8     6.5    0.28239          X          X X X X
       5    87.6     85.2     7.6    0.28816          X   X        X X X
       6    88.7     85.9     7.2    0.28079  X       X          X X X X
       6    88.6     85.9     7.3    0.28130          X   X      X X X X
       7    89.5     86.4     7.4    0.27600          X   X      X X X X X
       7    89.5     86.4     7.4    0.27622  X     X X          X X X X
       8    91.2     88.1     5.6    0.25799  X       X   X      X X X X X
       8    90.7     87.4     6.7    0.26525  X X     X          X X X X X
       9    91.5     88.0     6.9    0.25885  X X     X     X    X X X X X
       9    91.5     88.0     6.9    0.25897  X       X X X      X X X X X
      10    91.8     87.9     8.2    0.25996  X X     X X X      X X X X X
      10    91.7     87.7     8.5    0.26248    X X   X X X      X X X X X
      11    91.9     87.4    10.1    0.26577  X X X   X X X      X X X X X
      11    91.9     87.4    10.1    0.26587  X X   X X X X      X X X X X
      12    91.9     86.8    12.0    0.27228  X X X X X X X      X X X X X
      12    91.9     86.7    12.1    0.27261  X X X X X X        X X X X X
      13    91.9     86.0    14.0    0.27961  X X X X X X X X    X X X X X
```

  a. Using R_{adj}^2, the optimal model has 8 parameters ($R_{adj}^2 = 88.1\%$):
     $y = \beta_0 + \beta_1 x_2 + \beta_2 x_1^2 + \beta_3 x_3^2 + \beta_4 x_1 x_3 + \beta_5 x_1 x_4 + \beta_6 x_2 x_3 + \beta_7 x_2 x_4 + \beta_8 x_3 x_4 + \varepsilon$
  b. Using C_p, the optimal model has 12 parameters ($C_p = 12$):
     All parameters except x_4 and of course x_4^2.
  c. There is no good option in software to find the PRESS statistic for all subsets.
  d. The model with optimal C_p contains more parameters while the optimal R_{adj}^2 model is more parsimonious.

13.9 The Minitab output is given here:
Regression Analysis: y versus x1, x2, x1^2, x2^2, x1*x2

```
The regression equation is
y = 82 - 3.5 x1 + 0.97 x2 + 0.016 x1^2 - 0.0124 x2^2 + 0.053 x1*x2


Predictor      Coef    SE Coef       T      P
Constant       82.0      622.2    0.13  0.896
x1            -3.50      34.59   -0.10  0.920
x2            0.974      9.030    0.11  0.915
x1^2         0.0163     0.4864    0.03  0.974
x2^2       -0.01238    0.03687   -0.34  0.739
x1*x2        0.0527     0.2380    0.22  0.826

S = 22.4477    R-Sq = 61.4%    R-Sq(adj) = 54.7%

Analysis of Variance

Source          DF        SS        MS      F      P
Regression       5   23234.6    4646.9   9.22  0.000
Residual Error  29   14613.0     503.9
Total           34   37847.6

Source   DF    Seq SS
x1        1   14643.0
x2        1    6109.7
x1^2      1    1599.7
x2^2      1     857.5
x1*x2     1      24.7

Unusual Observations

Obs    x1      y     Fit  SE Fit  Residual  St Resid
  5  27.5  98.00   97.10   16.95      0.90      0.06 X
 35  16.3  97.00   96.84   17.47      0.16      0.01 X

X denotes an observation whose X value gives it large leverage.
```

a. $\hat{y} = 82.0 - 3.5x_1 + 0.974x_2 + 0.0163x_1^2 - 0.01238x_2^2 + 0.0527x_1x_2$.
b. The adjusted R^2 is not appreciably larger than the model with only x_2 and this full quadratic model has no significant terms (conditional on the others in the model).
c. Once again, I'd advocate removing variables, not adding them.

13.11 Minitab output is given here:
Regression Analysis: ln(y) versus x1, x2

```
The regression equation is
ln(y) = 3.08 - 0.0074 x1 + 0.0294 x2

Predictor       Coef    SE Coef       T      P
Constant       3.080      1.185    2.60  0.014
x1          -0.00743    0.03196   -0.23  0.818
x2          0.029416   0.009417    3.12  0.004

S = 0.525392   R-Sq = 51.4%   R-Sq(adj) = 48.4%

Analysis of Variance

Source           DF       SS      MS      F      P
Regression        2   9.3462  4.6731  16.93  0.000
Residual Error   32   8.8332  0.2760
Total            34  18.1794

Source  DF  Seq SS
x1       1  6.6526
x2       1  2.6936

Unusual Observations

Obs    x1   ln(y)     Fit  SE Fit  Residual  St Resid
 21  34.3  1.7918  3.0018  0.1805   -1.2100    -2.45R

R denotes an observation with a large standardized residual.
```

a. $\widehat{\ln(y)} = 3.08 - 0.0074x_1 + 0.0294x_2$
b. The R^2 and adjusted R^2 are both smaller in the log response model than in both the linear and quadratic models.
c. There is no reason we cannot compare the R^2_{adj} of the log response model to the standard model. However, we cannot compare the relative sizes of the model or residual sum of squares because they are on different scales.

13.13
a. Defining the "RoofType" variable in this fashion would indicate that this variable is quantitative not qualitative. A one-unit increase in RoofType could indicate a change from the cedar shingles to metal shingles, or a change from the metal shingles to asp[halt shingles, etc. There is no logical reason to assume that these two possible changes would indicate the same change in the response variable, y. Thus, the coefficient associated with this variable would be meaningless.

Chapter 13: Further Regression Topics

b. An improved approach would be to define these indicator variables:

$$x_1 = \begin{cases} 1 & \text{if RoofType=cedar shingles} \\ 0 & \text{if Otherwise} \end{cases}$$

$$x_2 = \begin{cases} 1 & \text{if RoofType=metal shingles} \\ 0 & \text{if Otherwise} \end{cases}$$

$$x_3 = \begin{cases} 1 & \text{if RoofType=asphalt shingles} \\ 0 & \text{if Otherwise} \end{cases}$$

13.15
a. We should include additional indicator variables for 'new' homes (age 0-10 years), 'mid' homes (age 11-20 years), and 'old' homes (age 21-30 years). We only need 2 of these so let's assume we include 'NEW' and 'MID'.
b. If the realtor's suspicions were true, the residuals positive for new homes (underestimate prices), near 0 for mid homes, and negative for old homes (overestimate prices).

13.17
a.

   a. Variety=Fuggle

Regression Analysis: P versus T, S

```
The regression equation is
P = 9.73 + 0.016 T - 0.759 S

Predictor      Coef   SE Coef       T       P
Constant      9.734     2.022    4.81   0.001
T            0.0159    0.1225    0.13   0.899
S           -0.7589    0.1804   -4.21   0.002

S = 0.611254   R-Sq = 68.6%   R-Sq(adj) = 62.3%
```

   b. Variety=Hallertau

Regression Analysis: P versus T, S

```
The regression equation is
P = 11.3 - 0.293 T - 0.263 S

Predictor      Coef   SE Coef       T       P
Constant     11.335     2.255    5.03   0.001
T           -0.2933    0.1478   -1.98   0.075
S           -0.2628    0.1961   -1.34   0.210

S = 0.645458   R-Sq = 58.4%   R-Sq(adj) = 50.1%
```

c. Variety=Northern Brewer
Regression Analysis: P versus T, S

```
The regression equation is
P = 20.9 - 0.277 T - 0.997 S

Predictor      Coef    SE Coef      T       P
Constant     20.890      5.047    4.14   0.002
T           -0.2773     0.2920   -0.95   0.365
S           -0.9965     0.3821   -2.61   0.026

S = 1.32296    R-Sq = 53.5%    R-Sq(adj) = 44.2%
```

d. Variety=Saaz
Regression Analysis: P versus T, S

```
The regression equation is
P = 13.9 - 0.343 T - 0.498 S

Predictor      Coef    SE Coef      T       P
Constant     13.851      2.920    4.74   0.001
T           -0.3425     0.1731   -1.98   0.076
S           -0.4981     0.1939   -2.57   0.028

S = 0.742204   R-Sq = 65.2%    R-Sq(adj) = 58.3%
```

b. **For Fuggle:** The residual plots versus the two independent variables do not indicate any higher level terms are needed. Using a full vs. reduced hypothesis test, it appears an interaction term is useful.

For Hallartau: The residual plots versus the two independent variables do not indicate any higher level terms are needed. Using a full vs. reduced hypothesis test, it appears an interaction term is not useful.

For Northern Brewer: The residual plots versus the two independent variables do not indicate any higher level terms are needed. Using a full vs. reduced hypothesis test, it appears an interaction term is not useful.

For Saaz: The residual plots versus the two independent variables do not indicate any higher level terms are needed. Using a full vs. reduced hypothesis test, it appears an interaction term is not useful.

For all of these models, determining the effectiveness of higher order terms is very difficult due to the same sample size. That is why there is no driving motivation to add additional terms in the residual plots.

13.19
a. The Minitab Output is on the next page
$$I = \begin{cases} 1 & \text{if} \quad \text{Variety=Saaz} \\ 0 & \text{if Variety=Hallertaur} \end{cases}$$

Regression Analysis: P versus T, S, I, I*T, I*S

```
The regression equation is
P = 11.3 - 0.293 T - 0.263 S + 2.52 I - 0.049 I*T - 0.235 I*S

Predictor      Coef    SE Coef       T       P
Constant     11.335      2.430    4.67   0.000
T           -0.2933     0.1592   -1.84   0.080
S           -0.2628     0.2113   -1.24   0.228
I             2.517      3.659    0.69   0.499
I*T         -0.0493     0.2273   -0.22   0.831
I*S         -0.2352     0.2786   -0.84   0.409

S = 0.695515    R-Sq = 62.6%   R-Sq(adj) = 53.3%
```

b. For Hallertau:
$$y = 11.3 - 0.293\,T - 0.263\,S$$

For Saaz:
$$y = 13.82 - 0.342\,T - 0.498\,S$$

c. 11.3 is the intercept for Hallertau
-0.293 is the partial slope for temperature for Hallertau
-0.263 is the partial slope for sunshine for Hallertau
11.3+2.52=13.82 is the intercept for Saaz
-0.293-0.049 = -0.342 is the partial slope for temperature for Saaz
-0.263-0.235 = -0.498 is the partial slope for sunshine for Saaz

d. For Hallertau:
$$y = 11.3 - 0.293\,(19) - 0.263\,(6.5) = 4.05$$

For Saaz:
$$y = 13.82 - 0.342\,(19) - 0.498\,(6.5) = 4.1$$

e. Hallertau
```
Predicted Values for New Observations

New
Obs    Fit   SE Fit      95% CI           95% PI
  1  4.054    0.215  (3.607, 4.502)   (2.536, 5.572)
```

Saaz
```
Predicted Values for New Observations

New
Obs    Fit   SE Fit      95% CI           95% PI
  1  4.106    0.224  (3.639, 4.573)   (2.582, 5.630)
```

13.21
  a. The estimated coefficient associated with PROMOTION is -19.960. This indicates that for fixed values of PRICE and CATEGORY, the average value of SALES is estimated to be reduced by 19.960 if a competing brand is having a promotion otherwise the average value of SALES does not change.
  b. One would suspect that a promotion by a truly competing brand would result in a decrease in sales. The model predicts this result since the estimated coefficient is negative.
  c. The t-statistic for testing whether the PROMOTION coefficient is different from 0 has $p-value < 0.0001$. Thus, there is significant evidence that the PROMOTION coefficient differs from 0.

13.23 When promotions are offered by a competing brand, PROMOTION=1, the model becomes:

$$\hat{y} = 26.807 + 90.233\ PRICE + 0.134\ CATEGORY + 287.609(1) - 142.433\ (PRICE)(1) - 0.024\ (CATEGORY)(0) \Rightarrow$$

$$\hat{y} = 314.416 - 52.200\ PRICE + 0.110\ CATEGORY$$

When promotions are not offered by a competing brand, PROMOTION=0, the model becomes:

$$\hat{y} = 26.807 + 90.233\ PRICE + 0.134\ CATEGORY + 287.609(0) - 142.433\ (PRICE)(0) - 0.024\ (CATEGORY)(0) \Rightarrow$$

$$\hat{y} = 26.807 + 90.233\ PRICE + 0.134\ CATEGORY$$

The model for predicting SALES have considerably different intercepts depending on whether or not there is a promotion for a competing brand. The partial slopes for PRICE for the two models have different signs and have very different magnitudes. The change in sign is of interest. It demonstrates that when there is a promotion for a competing brand, if the PRICE is increased, SALES drop considerably, whereas if there is not a promotion for a competing brand, a price increase does not result in a decrease in SALES.

13.25
  a. The Minitab Output is given here:
  Regression Analysis: consumpt versus price, income, temperat

```
The regression equation is
consumpt = 0.197 - 1.04 price + 0.00331 income + 0.00346 temperat

Predictor        Coef     SE Coef       T      P
Constant       0.1973      0.2702    0.73  0.472
price         -1.0444      0.8344   -1.25  0.222
income       0.003308    0.001171    2.82  0.009
temperat     0.0034584   0.0004455   7.76  0.000
```

```
S = 0.0368327   R-Sq = 71.9%   R-Sq(adj) = 68.7%


Analysis of Variance

Source            DF        SS        MS        F       P
Regression         3   0.090251  0.030084   22.17   0.000
Residual Error    26   0.035273  0.001357
Total             29   0.125523

Source      DF   Seq SS
price        1   0.008459
income       1   0.000051
temperat     1   0.081741
```

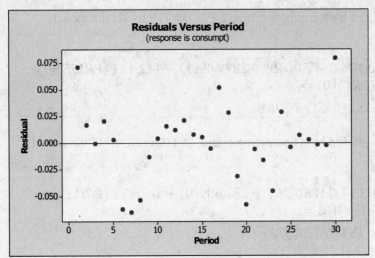

There appears to be some serial correlation in the errors.

b. The Durbin-Watson test statistic is 1.02117. Because the DW stat is less than 1.5, this implies a positive serial correlation.

13.27 The Minitab diagnostic plots are shown below

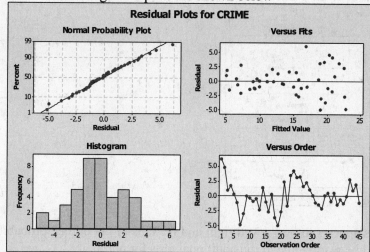

a. The histogram of residuals appears normal and the normal probability plot does not appear to deviation from a straight line which indicates that the assumption of normality does not appear to be violated.

b. The variance gets a little larger as the fit value increases. It would be worthwhile to run a test to verify the constant variance assumption.

c. $BP = \frac{SS(Regression)^*/2}{(\frac{SS(Residuals)}{45})^2} = \frac{740.89/2}{(257.09/45)^2} = 11.35 > 5.99 = \chi^2_{.05,2}$

Therefore, the BP stat rejects the assumption of constant variance.

d. The MSE results from regressions run on the transformed data are shown below:

λ	MSE
-1	35.1
-.5	15.54
0	8.42
.5	6.25
1	6.27
2	153.3

Because the λ yielding the lowest MSE is 0.5, the Box-Cox Transformation is a square root transformation. Because the MSE for λ=0.5 and λ=1 is so close, it may not be worth doing a transformation at all.

Chapter 13: Further Regression Topics

13.29 The model chosen is the model with x_2 only

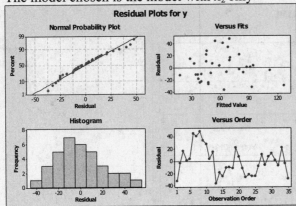

a. The histogram of residuals appears to be skewed right and the tails of the normal probability plot are a little askew. It would be worthwhile to run additional tests for normality.

b. The constant variance assumption does not appear to be violated.

c. $BP = \dfrac{SS(Regression)^*/2}{\left(\dfrac{SS(Residuals)}{34}\right)^2} = \dfrac{45714/2}{(17118/34)^2} = 0.09 > 3.84 = \chi^2_{.05,1}$

Therefore, the BP test does not reject the assumption of constant variance.

d. The MSE results from regressions run on the transformed data are shown below:

λ	MSE
-1	3495
-.5	1220
0	653
.5	510
1	519
2	882

Because the λ yielding the lowest MSE is 0.5, the Box-Cox Transformation is a square root transformation. Because the MSE for λ=0.5 and λ=1 is so close, it may not be worth doing a transformation at all.

13.31

a. $\hat{y} = -2.704 + 0.517\ RATES + 1.450\ UNEMPLOY + 0.0353\ RT5 * UNEP$

The fitted model has an $R^2 = 92.67\%$ and the three residual plots do not indicate any major pattern, thus the model appears to fit quite well.

b. A check of model conditions
1. Zero Expectation: The model appears to not need any higher order terms
2. Constant Variance: From the residuals vs predicted values, there does not appear to be an indication of unequal variance.
3. Normality: The boxplot appears slightly skewed to the right but there are no outliers. There is a slight indication of non-normality in the normal probability plots. Neither of these indications appears to require a transformation of the data.
4. The Durbin-Watson statistic equals 2.403 which would indicate a mild negative serial correlation but since it is less than 2.5, a differencing of the data is probably unnecessary.

13.33 The residual plot indicates that the model is underestimating y for small values of $\hat{y}$ and overestimating y for large values of $\hat{y}$. Thus, additional terms may be needed in the model. Since the data is quarterly earnings, there is a possibility of serial correlation. A plot of the residuals versus time would be recommended.

13.35
a.

Regression Analysis: Consumpt versus TempDiff, TempDiff^2, TempDiff^3

```
The regression equation is
Consumpt = - 15.6 + 24.2 TempDiff - 1.47 TempDiff^2 + 0.0291TempDiff^3

Predictor         Coef    SE Coef       T      P
Constant       -15.563      9.654   -1.61  0.119
TempDiff        24.197      2.910    8.31  0.000
TempDiff^2     -1.4748     0.2230   -6.61  0.000
TempDiff^3    0.029105   0.004722    6.16  0.000

S = 15.4227   R-Sq = 89.5%   R-Sq(adj) = 88.3%

Analysis of Variance

Source          DF      SS      MS      F      P
Regression       3   52941   17647  74.19  0.000
Residual Error  26    6184     238
  Lack of Fit    6    1408     235   0.98  0.463
  Pure Error    20    4776     239
Total           29   59125

Source         DF   Seq SS
TempDiff        1    41402
TempDiff^2      1     2503
TempDiff^3      1     9036


Unusual Observations

Obs  TempDiff  Consumpt     Fit  SE Fit  Residual  St Resid
 28      30.0    195.00  168.90    8.70     26.10     2.05R
R denotes an observation with a large standardized residual.

No evidence of lack of fit (P >= 0.1).
```

b. The LOF p-value given in the output is 0.483>0.05 which means there is no evidence of lack of fit.

c. The normality plots are give here:

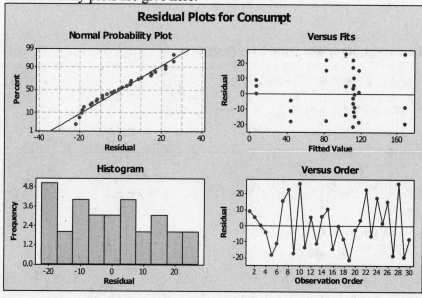

The normality probability plot and histogram shows non-normality in the residuals. The residual plot shows less variation at lower fitted value levels than at high.

13.37
   a.

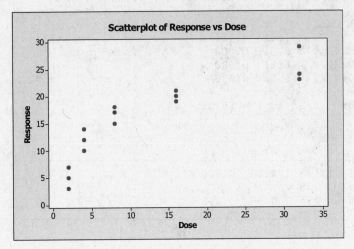

   b. Linear Model: $\hat{y} = 8.667 + 0.575\ DOSE$
      Quadratic Model: $\hat{y} = 4.484 + 1.506\ DOSE - 0.0270\ DOSE^2$

   c. The Quadratic Model appears to be more appropriate: it has a larger R^2 (88.15% vs. 77.30%), smaller MS(Error) (7.548 vs. 13.345), the term $DOSE^2$ has p-value=0.0062 which indicates that the quadratic term significantly improves the fit in comparison to the linear model, and the residuals are somewhat smaller in the quadratic model with less than an apparent pattern when compared to the residuals from the linear model.

   d. See SAS Output.

13.39
   a.

Correlations: bodywgt, liverwgt, reldose, proporti

```
            bodywgt    liverwgt    reldose
liverwgt     0.500
             0.029

reldose      0.990      0.490
             0.000      0.033

proporti     0.151      0.203      0.228
             0.537      0.404      0.349


Cell Contents:  Pearson correlation
                P-Value
```

There appears to be a large correlation between relative dose level and body weight. The correlation between body weight and liver weight as well as liver weight and relative dose are of mild concern.

b. The Minitab Output is given here:
Regression Analysis: proporti versus bodywgt, liverwgt, reldose

```
The regression equation is
proporti = 0.266 - 0.0212 bodywgt + 0.0143 liverwgt + 4.18 reldose

Predictor        Coef     SE Coef       T       P
Constant       0.2659      0.1946    1.37   0.192
bodywgt      -0.021246    0.007974  -2.66   0.018
liverwgt      0.01430     0.01722    0.83   0.419
reldose        4.178       1.523    2.74   0.015

S = 0.0772913    R-Sq = 36.4%    R-Sq(adj) = 23.7%

Analysis of Variance

Source           DF       SS         MS        F       P
Regression        3    0.051265   0.017088   2.86   0.072
Residual Error   15    0.089609   0.005974
Total            18    0.140874


Source      DF    Seq SS
bodywgt      1    0.003216
liverwgt     1    0.003067
reldose      1    0.044982

Unusual Observations

Obs   bodywgt   proporti     Fit    SE Fit   Residual   St Resid
  3     190      0.5600    0.5359   0.0713    0.0241      0.81 X

X denotes an observation whose X value gives it large leverage.
```

The p-value of the F-test is 0.072 suggests that additional terms are needed to fully explain the proportion of drug in the level.

c. Because of the large correlation between x_1 and x_3, only one of these variables should be included in the model. The best subsets regression is shown below:

Best Subsets Regression: proportion versus bodywgt, liverwgt, reldose

```
Response is proportion
                                       l
                                   b   i   r
                                   o   v   e
                                   d   e   l
                                   y   r   d
                                   w   w   o
                             Mallows  g   g   s
Vars  R-Sq  R-Sq(adj)    Cp       S   t   t   e
  1    5.2        0.0   7.4   0.088643        X
  1    4.1        0.0   7.6   0.089130   X
  2   33.5       25.1   2.7   0.076538   X    X
  2    6.3        0.0   9.1   0.090835   X X
  3   36.4       23.7   4.0   0.077291   X X  X
```

The highest R^2 adjusted comes from the model with body weight and relative dose.

13.41
a. A scatterplot of the data is given here:

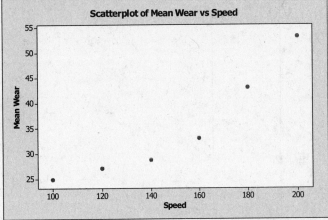

It would appear that a quadratic model in Machine Speed is needed:
$$Wear = \beta_0 + \beta_1 Speed + \beta_2 Speed^2 + \varepsilon$$

b. Minitab Output is given here

Regression Analysis: Wear versus Speed, speed^2

```
The regression equation is
Wear = 63.1 - 0.705 Speed + 0.00328 speed^2

Predictor        Coef    SE Coef       T      P
Constant       63.139      6.125   10.31  0.000
Speed         -0.70507    0.08469   -8.33  0.000
speed^2       0.0032768  0.0002810  11.66  0.000
S = 1.94220    R-Sq = 96.6%    R-Sq(adj) = 96.5%
```

```
Analysis of Variance

Source           DF      SS      MS       F       P
Regression        2   4839.9  2419.9   641.53   0.000
Residual Error   45    169.7     3.8
Total            47   5009.6

Source    DF   Seq SS
Speed      1   4326.8
speed^2    1    513.1

Unusual Observations

Obs  Speed   Wear    Fit   SE Fit  Residual  St Resid
 42    200  43.700  53.196  0.622   -9.496    -5.16R

R denotes an observation with a large standardized residual.
```

The estimated regression equation is
$\hat{y} = 63.139 - 0.70507\, x_1 + 0.0032768 x_1^2$

c. A residual plot for the fitted model is given here:

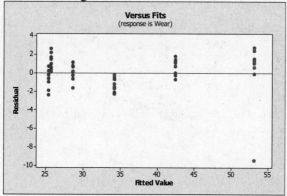

It would appear that the model is not an adequate representation of the variation in Wear since at some Machine Speeds all the residuals are positive and at other Machine Speeds all the residuals are negative. Although the model overall is providing an excellent fit to the data, this pattern would indicate that further modeling is needed. For example, there may be other independent variables beside Machine Speed which may affect Wear.

13.43
Regression Analysis: Wear versus x1, x1^2, x2

```
The regression equation is
Wear = 60.5 - 0.705 x1 + 0.00328 x1^2 + 8.88 x2

Predictor       Coef     SE Coef      T       P
Constant       60.477     5.512    10.97   0.000
x1            -0.70507    0.07551   -9.34   0.000
x1^2          0.0032768  0.0002505  13.08   0.000
x2             8.875      2.499     3.55   0.001

S = 1.73165    R-Sq = 97.4%   R-Sq(adj) = 97.2%
```

Analysis of Variance

```
Source          DF      SS       MS       F        P
Regression       3    4877.7   1625.9   542.22   0.000
Residual Error  44     131.9      3.0
Total           47    5009.6
Source          DF   Seq SS
x1               1   4326.8
x1^2             1    513.1
x2               1     37.8
```

Unusual Observations

```
Obs    x1    Wear    Fit    SE Fit   Residual   St Resid
42    200   43.700  52.309   0.609    -8.609     -5.31R
```

The first fitted regression equation is: $\hat{y} = 60.477 - 0.705x_1 + 0.00328x_1^2 + 8.875x_2$

Regression Analysis: Wear versus x1, x1^2, x2, x2^2, x1*x2

* x2^2 is highly correlated with other X variables
* x2^2 has been removed from the equation.

The regression equation is
Wear = 64.4 - 0.731 x1 + 0.00328 x1^2 - 4.0 x2 + 0.0861 x1*x2

```
Predictor      Coef      SE Coef     T        P
Constant      64.350      6.392    10.07    0.000
x1           -0.73089     0.07828  -9.34    0.000
x1^2          0.0032768   0.0002494 13.14   0.000
x2           -4.04       11.21     -0.36    0.721
x1*x2         0.08607     0.07285   1.18    0.244
```
S = 1.72391   R-Sq = 97.4%   R-Sq(adj) = 97.2%

Analysis of Variance

```
Source          DF      SS       MS       F        P
Regression       4    4881.8   1220.5   410.67   0.000
Residual Error  43     127.8      3.0
Total           47    5009.6
```

```
Source          DF   Seq SS
x1               1   4326.8
x1^2             1    513.1
x2               1     37.8
x1*x2            1      4.1
```

Unusual Observations

```
Obs    x1    Wear    Fit    SE Fit   Residual   St Resid
42    200   43.700  51.879   0.707    -8.179     -5.20R
```

The second fitted regression equation is:
$$\hat{y} = 42.28 - 0.421x_1 + 0.00224x_1^2 + 69.54x_2 - 0.949x_1x_2 + 0.00345x_1^2x_2$$

These two models provide only marginal improvement over the quadratic model in just x_1. However, the pattern in the residual plot noted from the quadratic model in x_1 is not as noticeable in the residual plots from the two models.

A residual plot of the first fitted model is given here:

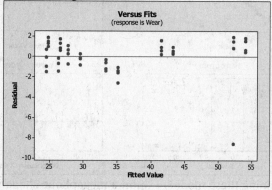

13.45   There is no indication in the plot of Height by Amount of a quadratic curvature in the relationship. Hence, the second order terms in Amount are probably unnecessary.

13.47
Best Subsets Regression: EHg versus Alkalinity, pH, ...

```
Response is EHg
                                   A
                                   l
                                   k
                                   a         C
                                   l    a C                      c p
                                   i    l h       p         a h H
                                   n    c l a   H p     a   * * *
                                   i    i o * * * c c c
                          Mallows  t    p u r c c p a a a
Vars  R-Sq  R-Sq(adj)   Cp      S  y    H m o h h H l l l
  1   39.5     38.3    8.7  0.26598                     X
  1   39.4     38.2    8.8  0.26619  X
  2   45.0     42.8    5.5  0.25627  X      X
  2   44.8     42.6    5.6  0.25661  X X
  3   48.6     45.4    4.1  0.25023  X      X           X
  3   48.5     45.3    4.2  0.25051  X      X X
  4   54.0     50.2    0.9  0.23902  X      X           X X
  4   53.3     49.4    1.5  0.24089  X            X X X
  5   55.0     50.2    1.9  0.23901  X      X     X X X
  5   54.9     50.1    2.0  0.23918  X X    X           X X
  6   55.6     49.8    3.4  0.23995  X X    X     X X X
  6   55.2     49.4    3.7  0.24101  X      X X   X X X
  7   55.9     49.0    5.1  0.24194  X X    X X X X X
  7   55.8     48.9    5.2  0.24214  X X    X     X X X X
  8   55.9     47.9    7.0  0.24443  X X    X X X X X X
  8   55.9     47.9    7.1  0.24449  X X X  X X X X X
  9   56.0     46.7    9.0  0.24719  X X X  X X X X X X
  9   56.0     46.7    9.0  0.24721  X X    X X X X X X X
 10   56.0     45.5   11.0  0.25007  X X X  X X X X X X X
```

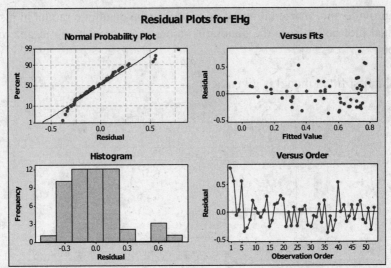

The residuals found when running these models exhibit a megaphone shape which indicates non-constant variance. To counter this, we will model the ln(EHg).

Best Subsets Regression: lnEHg versus Alkalinity, pH, ...

```
Response is lnEHg
                                            A
                                            l
                                            k
                                            a         C
                                            l    a C         c p
                                            i    l h   p   a h H
                                            n    c l a H a * * *
                                            i    i o * * * c c c
                              Mallows       t    p u r c c p a a a
Vars  R-Sq  R-Sq(adj)   Cp       S          y    H m o h h H l l l
  1   59.2     58.4    34.4   0.52696                  X
  1   57.7     56.9    37.4   0.53649                              X
  2   74.5     73.5     5.1   0.42079       X    X
  2   74.2     73.1     5.8   0.42337       X         X
  3   75.8     74.3     4.5   0.41421       X    X                 X
  3   75.5     74.0     5.1   0.41670       X    X X
  4   77.6     75.8     2.6   0.40193       X    X       X X
  4   77.6     75.7     2.8   0.40254       X            X X       X
  5   78.3     76.0     3.3   0.40034       X    X       X   X X
  5   78.3     76.0     3.4   0.40043       X            X X X X
  6   79.1     76.4     3.7   0.39693       X X          X X X X
  6   79.1     76.3     3.7   0.39730       X X  X       X   X X
  7   79.3     76.1     5.3   0.39955       X X      X X X   X X
  7   79.3     76.1     5.3   0.39963       X X  X       X X X X
  8   79.4     75.6     7.1   0.40332       X        X X X X X X
  8   79.3     75.6     7.2   0.40356       X X X    X X X   X X
  9   79.4     75.1     9.0   0.40734       X X X    X X X X X X
  9   79.4     75.1     9.1   0.40779       X X X X      X X X X
 10   79.4     74.5    11.0   0.41216       X X X X X X X X X X
```

a. Using R^2_{adj} as the guideline, it appears a model including alkalinity, pH. pH*chloro, alkalinity*pH, chloro*cal, and pH*calcium is the best model.

Regression Analysis: lnEHg versus Alkalinity, pH, ...

The regression equation is
lnEHg = 0.377 - 0.0541 Alkalinity - 0.109 pH - 0.00104 pH*ch +
        0.00504 a*pH - 0.000215 ch*cal + 0.00232 pH*cal

```
Predictor         Coef     SE Coef       T      P
Constant        0.3767      0.4537    0.83  0.411
Alkalinity    -0.05410     0.01583   -3.42  0.001
pH            -0.10868     0.08032   -1.35  0.183
pH*ch        -0.0010411   0.0006913  -1.51  0.139
a*pH          0.005041    0.001958    2.57  0.013
ch*cal       -0.0002154   0.0001256  -1.72  0.093
pH*cal        0.0023167   0.0008209   2.82  0.007

S = 0.396933   R-Sq = 79.1%   R-Sq(adj) = 76.4%
```

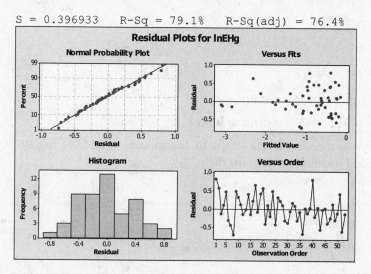

b. The model conditions appear to be valid for the ln(EHg) model, but not for the EHg prediction model (non-constant variance and non-normality of residuals).

c. $\ln(EHg) = 0.3767 - 0.05410(80) - 0.10868(6) - 0.0010411(6*40) + 0.005041(6*80) - 0.0002154(60*40) + 0.0023167(60*6) = -2.116512$
Therefore, the predicted EHg is $e^{-2.116512} = 0.12045$

13.49 Minitab output is given here:

Regression Analysis: Solubili versus Temp, Temp^2

The regression equation is
Solubili = 44.2 - 0.494 Temp + 0.00143 Temp^2

```
Predictor        Coef     SE Coef      T      P
Constant       44.182       1.756  25.16  0.000
Temp          -0.49400     0.06688  -7.39  0.000
Temp^2        0.0014285   0.0005163   2.77  0.017

S = 3.27314   R-Sq = 96.0%   R-Sq(adj) = 95.3%
```

Chapter 13: Further Regression Topics

```
Analysis of Variance

Source           DF      SS      MS       F       P
Regression        2   3071.2  1535.6  143.33   0.000
Residual Error   12    128.6    10.7
  Lack of Fit     3     36.7    12.2    1.20   0.364
  Pure Error      9     91.8    10.2
Total            14   3199.7
Source   DF  Seq SS
Temp      1  2989.2
Temp^2    1    82.0

Unusual Observations
Obs  Temp  Solubili   Fit    SE Fit  Residual  St Resid
 10    75    9.000  15.168   1.285    -6.168    -2.05R

R denotes an observation with a large standardized residual.

No evidence of lack of fit (P >= 0.1).
```

a. The fitted model is $\hat{y} = 44.182 - 0.494x + 0.00143x^2$

b. From the output, $F = \frac{36.7/3}{91.8/9} = 1.20 \Rightarrow p-value = 0.364$

Thus, there is not significant evidence of lack of fit of the model. Thus the higher terms in Temperature (x) are not needed to adequately fit the data.

c. A residual plot of the fitted data is given here:

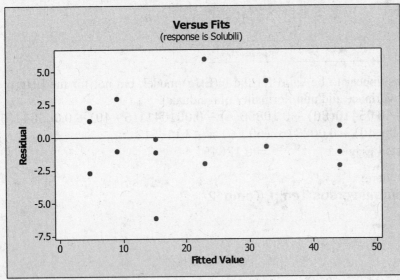

There are no obvious patterns in the residual plot.

13.51 The calculations for the test of lack of fit are given here:

x (Dose Level)	$\bar{y}_i$	$\sum_j (y_{ij} - \bar{y}_i)^2$	$n_i - 1$
2	5	8	2
4	12	8	2
8	16.667	4.667	2
16	20	2	2
32	25.333	20.667	2
Total		43.334	10

$SSP_{exp} = 43.334 \quad df_{exp} = 10$

From the output from Exercise 13.37, SS(residual)=90.579 $df_{residual}$=12
The SS_{Lack}=90.579−43.334=47.245 df_{Lack}=12−10=2
$$F = \frac{47.245/2}{43.334/10} = 5.45 \quad df = 2, 10 \Rightarrow p-value = 0.0251$$
There is significant evidence of lack of fit of the quadratic model. Hence, higher order terms in Dose Level, such as x^3, x^4 may be required to improve the fit of the model.

13.53
a. A scatterplot of the data is given in the textbook.
b. The estimated linear regression equation is given here:
$\hat{y} = -1.540 + 0.70635x$

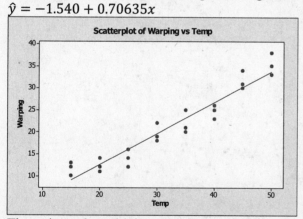

c. The estimated quadratic regression equation is given here:
$\hat{y} = 9.179 - 0.0468x + 0.011587x^2$

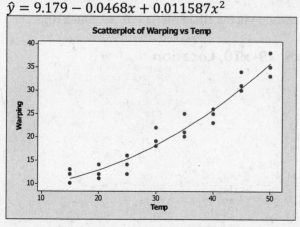

174   Chapter 13: Further Regression Topics

d. Using Linear equation, then Temperature is 27°C, $\hat{y} = 17.5$
Using Quadratic equation, then Temperature is 27°C, $\hat{y} = 16.4$

13.55
a. The correlation matrix is shown below with 'large' correlations (>0.8) bolded.

Correlation: x1, x2, x3, x4, x5, x6, x7, x8, x9, x10

	x1	x2	x3	x4	x5	x6	x7	x8	x9
x2	**0.842**								
x3	**0.938**	**0.933**							
x4	**0.896**	**0.840**	**0.914**						
x5	0.449	0.681	0.594	0.570					
x6	**0.808**	**0.818**	**0.870**	**0.874**	0.783				
x7	-0.179	-0.166	-0.161	-0.103	-0.122	-0.042			
x8	-0.687	-0.340	-0.533	-0.527	0.191	-0.302	0.172		
x9	-0.758	-0.482	-0.679	-0.655	-0.065	-0.537	0.272	**0.911**	
x10	-0.089	0.030	-0.093	-0.090	0.411	0.128	-0.147	0.346	0.223

These high correlations can make the regression results unstable.

b. There are no major violations to the assumptions of normality or equal variance, but there are a few residuals on the low end to be aware of.

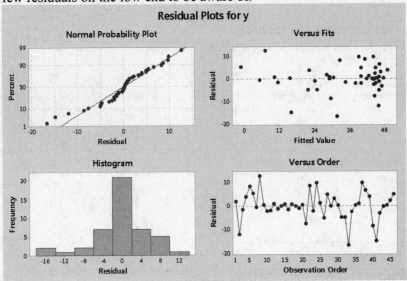

13.57 Results here will be based on the model chosen via the BIC criteria (x6, x9, and x10).
a. The regression output is shown below.

Regression Analysis: y versus x6, x9, x10, Location

```
Method

Categorical predictor coding   (1, 0)

Analysis of Variance

Source              DF    Adj SS    Adj MS   F-Value   P-Value
Regression           7   8344.00   1192.00     34.89     0.000
  x6                 1    330.20    330.20      9.67     0.004
  x9                 1   1309.14   1309.14     38.32     0.000
```

```
x10              1    172.16   172.16   5.04   0.031
Location         1     88.85    88.85   2.60   0.115
x6*Location     1      1.66     1.66   0.05   0.827
x9*Location     1    267.65   267.65   7.84   0.008
x10*Location    1     32.01    32.01   0.94   0.339
Error           38  1298.11    34.16
Total           45  9642.11
```

Model Summary

```
    S     R-sq   R-sq(adj)  R-sq(pred)
5.84473   86.54%    84.06%     79.24%
```

Coefficients

```
Term          Coef    SE Coef   T-Value   P-Value    VIF
Constant      96.5    26.5      3.64      0.001
x6            0.2336  0.0751    3.11      0.004     3.27
x9           -0.2803  0.0453   -6.19      0.000     2.35
x10           0.01939 0.00864   2.24      0.031     2.18
Location
  W           84.4    52.3      1.61      0.115    921.03
x6*Location
  W           0.032   0.147     0.22      0.827    299.98
x9*Location
  W          -0.2291  0.0818   -2.80      0.008    336.88
x10*Location
  W          -0.0142  0.0147   -0.97      0.339      6.14
```

Regression Equation

```
Location
E        y = 96.5 + 0.2336 x6 - 0.2803 x9 + 0.01939 x10

W        y = 180.8 + 0.266 x6 - 0.5093 x9 + 0.0052 x10
```

Fits and Diagnostics for Unusual Observations

```
Obs    y      Fit    Resid   Std Resid
 22  50.00  37.28   12.72     2.33  R
 24  54.00  39.79   14.21     2.62  R
 33  15.00  30.05  -15.05    -2.83  R
```

Using the partial F test, we will test if the Location and the interaction terms are significant.

$$F = \frac{(8344 - 7762.2)/4}{1298.11/38} = 4.25 \Rightarrow p-value = 0.0061$$

Therefore, there is significant evidence that the location is a significant factor in predicting daily amount of evaporation of the soil.

b.

Location	Fit	95% CI
East	42.15	(36.999, 47.297)
West	40.06	(32.0154, 48.108)

c. There is little difference in the average fit but the West is lower. The interval is wider in the West.

13.59 Test the hypothesis, $H_0: \beta_{AGE} \geq -2500$ vs. $H_a: \beta_{AGE} < -2500$

$t = \frac{\hat{\beta}_{AGE} - 2500}{SE(\hat{\beta}_{AGE})} = \frac{-506 - (-2500)}{1111} = 1.79$ with df=12

$p - value = \Pr(t_{12} \leq 1.79) = 0.9507 \Rightarrow$

Fail to reject H_0. There is not sufficient evidence that the deprecation per year is less than $2500.

13.61 The following table summarizes the fit of many models. The p-values for comparing the reduced model to the full model (model listed directly above it) are also given along with the maximum p-value for testing whether to drop any additional term in the given model.

Number of Ind. Var in the Model	R^2_{adj}	MS(Residual)	df_{Res}	p-value for reducing Model	Var Removed from Model
9	47.2%	273282828	12	0.898	BEDB,AGE
7	53.9%	238486598	14	0.412	BEDA
6	54.8%	233974343	15	0.346	BATHS
5	54.9%	233174830	16	0.136	CARB
4	51.1%	253230451	17	0.188	LOT
3	48.7%	265619875	18	0.081	DOM
2	42.1%	299566628	19	0.013	CARA
1	23.0%	398320115	20	0.014	BEDC

The "best" model is the model with the independent variables: DOM, CARA, and BEDC. It has considerably larger value for R^2_{adj} than the two-variable model without DOM, the p-value for dropping DOM from this model is only 0.081, which is marginally significant, it has a $C_p=3.5$ which is fairly close to p=3 and there is a considerable increase in MS(Residual) when DOM is dropped from the model.

13.63
a. The question is a test of $H_0: \beta_1 = \beta_2 = 0$ vs $H_a: \beta_1 \neq 0$ and/or $\beta_2 \neq 0$.
From the output, $F = \frac{MS(Model)}{MS(Error)} = 15.987$, with p-value $< 0.0001 < 0.05 \Rightarrow$

Reject H_0 and conclude there is significant evidence that ROOMS and SQFT taken together contain information about PRICE.

b. Test $H_0: \beta_1 = 0$ vs $H_a: \beta_1 \neq 0$
t = 0.717 with p-value=0.4822 > 0.05 $\Rightarrow$
Fail to reject H_0 and conclude there is not significant evidence that the coefficient of ROOMS is different from 0.

c. Test $H_0: \beta_2 = 0$ vs $H_a: \beta_2 \neq 0$
t = 1.468 with p-value=0.1585 > 0.05 $\Rightarrow$
Fail to reject H_0 and conclude there is not significant evidence that the coefficient of SQFT is different from 0.

13.65   The F-test for the overall model is 4.42 with p-value = 0.0041.
The indicator variable DV_3 measures the difference in risk of infection between hospital in the South and West holding all other variables constant. The coefficient of DV_3 is β_7 and we want to test $H_0: \beta_7 \leq .5\%$ vs $H_a: \beta_7 > .5\%$. The test statistic is:

$$t = \frac{\hat{\beta}_7 - 0.5}{SE(\hat{\beta}_7)} = \frac{0.7024 - 0.5}{0.8896} = 0.23 \text{ with df=20}$$

$p - value = \Pr(t_{20} > 0.23) = 0.4102 \Rightarrow$
Fail to reject H_0, there is not significant evidence that the infection rate in the south is at least 0.5% higher than in the west.

13.67   The table on page 870 provides summary information for a one variable at a time elimination from the full model. The following model is selected based on this information:
$y = \beta_0 + \beta_1 STAY + \beta_3 RCR + \epsilon$
The R^2 for this model is 0.5578 vs. 0.6072 for the seven variable model.
The MS(Error) for this model is 28.765 vs. 25.546 for the seven variable model.
A test of H_0: Two Variable Model vs. H_a: Seven Variable model is given by testing the following parameters in the seven variable model:
$H_0: \beta_2 = \beta_4 = \beta_5 = \beta_6 = \beta_7 = 0$ vs H_a: at least one of $\beta_2, \beta_4, \beta_5, \beta_6, \beta_7 \neq 0$
$$F = \frac{(39.49805177 - 36.27961297)/5}{25.54623394/20} = 0.50 \text{ with } df = 5,20 \Rightarrow$$
$p - value = \Pr(F_{5,20} > 0.50) = 0.7726 \Rightarrow$
Fail to reject H_0, there is not significant evidence that at least one of the five parameters is not 0. Thus, there is not significant evidence of a difference between the two-variables and seven-variables models.
Based on the above test, the marginal difference in R^2 and MS(Error), the model with fewer variables is the more desirable model.

13.69
   a.   The model F-test has value 6.52 with p-value=0.0004. Thus we can reject $H_0: \beta_1 = \beta_2 = \beta_3 = \beta_4 = \beta_5 = \beta_6 = 0$ and conclude that at least one of $\beta_1, \beta_2, \beta_3, \beta_4, \beta_5, \beta_6 \neq 0$. There is significant evidence that the set of six independent variables does provide an explanation of the variability in PULSE.
   b.   The pairwise correlations between the six independent variables ranged from -0.74863 to 0.59885. Thus, there may be a degree of multicollinearity due to pairwise correlations. The largest correlation was between PHYS1 and PHYS2. Multicollinearity may greatly inflate the standard errors of the predictions.
   c.   The coefficient of the variables PHYS1 measures the difference in average pulse between individuals who have substantial physical exercise and those who exercise little. The estimate of this parameter is 13.43 which indicates individuals who exercise substantially have a mean increase in pulse rate of 13.43 units higher than those who exercise very little.
A 95% C.I. for the coefficient of PHYS1 is $13.43 \pm (t_{.025,23})(SE(\beta)) \Rightarrow$
$13.43 \pm (2.069)(4.2512) \Rightarrow (4.63, 22.23)$
Thus, we are 95% confident that the actual average difference in the increase in pulse between the two groups is greater than zero.

178 Chapter 13: Further Regression Topics

13.71
a. Based on the information contained in the model fit summary table, dropping HEIGHT and WEIGHT from the original six variable model did not substantially affect the fit of the model.
The six variable model has $C_p = 7 > 6 = p$, whereas, the four-variable model has $C_p = 3.26 \approx 4 = p$
The Six-variable model has MS(Error)=47.3, which is larger than the value for the four-variable model: MS(Error)=44.02
The Six-variable model has R^2=0.6297, which is only slightly larger than the value for the four-variable model: R^2=0.625
The best model amongst models having similar values for the criteria used to evaluate the fit of the model is the model having the fewest independent variables. Thus, the four-variable model, excluding HEIGHT and WEIGHT, is the best model.

b. The model with interactions between all qualitative variables would be:
$$y = \beta_0 + \beta_1 RUN + \beta_2 SMOKE + \beta_3 PHYS1 + \beta_4 PHYS2 + \beta_5 RUN * SMOKE + \beta_6 RUN * PHYS1 + \beta_7 RUN * PHYS2 + \beta_8 SMOKE * PHYS1 + \beta_9 SMOKE * PHYS2 + \epsilon$$
There is no term for PHYS1*PHYS2 since there is no meaning for such a term.

13.73
a. No, the linear regression lines will not change because they are based on the mean of the y-values.
b. Predictions of x will change because we are now predicting means of four measurements, rather than individual values.

13.75
a. The following Minitab output contains the equations
Regression Analysis: Measurem versus I1, I2, I3, x, x*I1, x*I2, x*I3

```
The regression equation is
Measurem = 30.4 + 23.5 I1 - 10.8 I2 - 11.0 I3 + 543 x - 35.6 x*I1 - 94.7 x*I2 -
69.2 x*I3

Predictor       Coef    SE Coef        T       P
Constant      30.412      7.220     4.21   0.000
I1             23.51      10.21     2.30   0.024
I2            -10.77      10.21    -1.06   0.295
I3            -10.97      10.21    -1.07   0.286
x            542.875      2.843   190.95   0.000
x*I1         -35.625      4.021    -8.86   0.000
x*I2         -94.750      4.021   -23.57   0.000
x*I3         -69.250      4.021   -17.22   0.000

S = 5.68595   R-Sq = 100.0%   R-Sq(adj) = 99.9%

Analysis of Variance

Source           DF        SS         MS        F       P
Regression        7   4662476     666068  20602.14   0.000
Residual Error   72      2328         32
Total            79   4664804
```

b. The separate regression equations are:
Lab 1: $\hat{y} = (30.413 + 23.51) + (542.875 - 35.625)x \Rightarrow y = 53.923 + 507.25x$
Lab 2: $\hat{y} = (30.413 - 10.77) + (542.875 - 94.750)x \Rightarrow y = 19.643 + 448.125x$
Lab 3: $\hat{y} = (30.413 - 10.97) + (542.875 - 69.250)x \Rightarrow y = 19.443 + 473.625x$
Lab 4: $\hat{y} = 30.413 + 542.875x \Rightarrow y = 30.413 + 542.875x$

c. The models are identical except for round-off error in computing the intercepts.

d. The single model has a considerably larger value for degrees of freedom for error and hence more accurate tests of hypotheses and confidence intervals can be obtained from this model than from the individual models. Also, the single model permits the testing whether or not there is a significant difference in the intercepts and slopes between the four labs.

13.77

a.

Best Subsets Regression: y versus x_1, x_2, x_3, x_4, x_5, x_6
Response is y

Vars	R-Sq	R-Sq(adj)	Mallows Cp	S	x_1	x_2	x_3	x_4	x_5	x_6
1	41.6	40.1	23.1	18.170	X					
1	24.4	22.4	40.8	20.671		X				
1	18.8	16.7	46.5	21.420	X					
1	13.7	11.4	51.8	22.088					X	
2	58.6	56.5	7.6	15.489		X	X			
2	51.6	49.1	14.8	16.752	X	X				
2	49.8	47.2	16.6	17.060	X				X	
2	42.1	39.1	24.5	18.318	X			X		
3	61.7	58.6	6.4	15.096		X	X		X	
3	61.3	58.1	6.9	15.191	X	X	X			
3	59.3	56.0	8.9	15.569		X	X	X		
3	59.3	56.0	8.9	15.570			X	X	X	
4	64.0	60.0	6.1	14.853	X	X	X		X	
4	63.3	59.2	6.8	14.991	X	X	X	X		
4	62.9	58.8	7.2	15.068	X	X	X			X
4	62.8	58.7	7.2	15.081		X	X	X		X
5	66.9	62.1	5.1	14.447	X	X	X	X	X	
5	65.0	60.0	7.0	14.843	X	X	X	X		X
5	64.0	58.8	8.1	15.063	X	X	X		X	X
5	62.9	57.6	9.2	15.284		X	X	X	X	X
6	67.0	61.1	7.0	14.636	X	X	X	X	X	X

b. The model with 5 parameters (all but x_6) with $R^2_{adj}=62.1\%$ gives the highest R^2_{adj} and a C_p closest to the number of parameters.

c. The variables chosen in the five variable model are number of manufacturing enterprises employing 20 or more workers, population size, average annual temperature, average annual wind speed, and average annual precipitation.

180   Chapter 13: Further Regression Topics

13.79   The Minitab output of the five-variable model is shown below with the log transformed data
Regression Analysis: ln(y) versus x_1, x_2, x_3, x_4, x_5

```
The regression equation is
ln(y) = 7.35 - 0.0611 x_1 + 0.00126 x_2 - 0.000707 x_3 - 0.171 x_4 + 0.0181 x_5

Predictor          Coef     SE Coef        T      P      VIF
Constant         7.3499      0.9254     7.94  0.000
x_1             -0.06113     0.01271    -4.81  0.000    1.731
x_2            0.0012642    0.0004751    2.66  0.012   14.703
x_3           -0.0007072    0.0004566   -1.55  0.130   14.339
x_4             -0.17054     0.05397    -3.16  0.003    1.219
x_5             0.018148     0.006610    2.75  0.009    1.242

S = 0.441601    R-Sq = 65.4%    R-Sq(adj) = 60.5%


Analysis of Variance

Source           DF        SS         MS       F       P
Regression        5    12.9035     2.5807   13.23   0.000
Residual Error   35     6.8254     0.1950
Total            40    19.7289


Source    DF    Seq SS
x_1        1    5.8914
x_2        1    2.9631
x_3        1    1.0818
x_4        1    1.4971
x_5        1    1.4701

Unusual Observations

Obs    x_1    ln(y)     Fit    SE Fit   Residual   St Resid
  1   70.3   2.3026   2.0151   0.3101    0.2875       0.91 X
 11   50.6   4.7005   4.9530   0.3677   -0.2526      -1.03 X
 31   50.0   4.5433   3.5687   0.1601    0.9746       2.37 R

R denotes an observation with a large standardized residual.
X denotes an observation whose X value gives it large leverage.
```

    a.   There are 2 points (1 and 11) with high leverage values and one (31) with a large residual.

        Minitab output without high leverage points included
        Regression Analysis: ln(y) versus x_1, x_2, x_3, x_4, x_5

```
The regression equation is
ln(y) = 7.39 - 0.0625 x_1 + 0.00127 x_2 - 0.000507 x_3 - 0.184 x_4 + 0.0188 x_5

    Predictor          Coef     SE Coef        T      P      VIF
    Constant         7.3875      0.8695     8.50  0.000
    x_1             -0.06250     0.01341    -4.66  0.000    1.909
    x_2            0.0012716    0.0005164    2.46  0.019    6.471
    x_3           -0.0005069    0.0004482   -1.13  0.266    6.412
    x_4             -0.18391     0.05402    -3.40  0.002    1.149
    x_5             0.018845     0.007806    2.41  0.022    1.636
```

```
S = 0.413942   R-Sq = 62.4%   R-Sq(adj) = 56.5%

Analysis of Variance

Source           DF      SS       MS      F      P
Regression        5   9.0803   1.8161  10.60  0.000
Residual Error   32   5.4831   0.1713
Total            37  14.5635

Source  DF   Seq SS
x_1      1   3.7697
x_2      1   1.4375
x_3      1   0.6211
x_4      1   2.2533
x_5      1   0.9987


Unusual Observations

Obs   x_1    ln(y)    Fit    SE Fit   Residual   St Resid
 34   59.3   3.4340  2.5400  0.1319    0.8940      2.28R

R denotes an observation with a large standardized residual.
```

13.81

a. The model is shown below.

Regression Analysis: Math versus Reading, %Minority, %Poverty, Grade

```
Method

Categorical predictor coding   (1, 0)
Rows unused                       3


Analysis of Variance

Source            DF    Adj SS   Adj MS   F-Value  P-Value
Regression        11   17368.6  1578.97   127.44    0.000
  Reading          1     153.5   153.50    12.39    0.001
  %Minority        1      35.3    35.29     2.85    0.098
  %Poverty         1       0.8     0.79     0.06    0.802
  Grade            2      28.0    13.99     1.13    0.331
  Reading*Grade    2      18.1     9.04     0.73    0.487
  %Minority*Grade  2      50.3    25.17     2.03    0.142
  %Poverty*Grade   2     108.8    54.39     4.39    0.017
Error             51     631.9    12.39
Total             62   18000.5
```

Model Summary

S	R-sq	R-sq(adj)	R-sq(pred)
3.51996	96.49%	95.73%	94.20%

Coefficients

Term	Coef	SE Coef	T-Value	P-Value	VIF
Constant	17.1	48.7	0.35	0.727	
Reading	0.920	0.261	3.52	0.001	96.60
%Minority	-0.0925	0.0548	-1.69	0.098	9.28
%Poverty	0.0178	0.0705	0.25	0.802	14.10
Grade					
4	68.6	70.9	0.97	0.338	5808.60
5	83.8	55.9	1.50	0.140	3344.09
Reading*Grade					
4	-0.321	0.361	-0.89	0.378	5230.54
5	-0.346	0.288	-1.20	0.235	3770.75
%Minority*Grade					
4	0.0370	0.0847	0.44	0.664	20.52
5	0.1576	0.0805	1.96	0.056	17.87
%Poverty*Grade					
4	-0.108	0.108	-1.00	0.321	50.50
5	-0.2867	0.0978	-2.93	0.005	40.33

Regression Equation

Grade

3   Math = 17.1 + 0.920 Reading - 0.0925 %Minority + 0.0178 %Poverty

4   Math = 85.7 + 0.599 Reading - 0.0555 %Minority - 0.0901 %Poverty

5   Math = 100.9 + 0.574 Reading + 0.0652 %Minority - 0.2689 %Poverty

Fits and Diagnostics for Unusual Observations

Obs	Math	Fit	Resid	Std Resid	
6	169.80	163.05	6.75	2.11	R
32	194.50	191.11	3.39	1.57	X
45	197.10	188.25	8.85	2.85	R
53	210.20	211.50	-1.30	-0.69	X

R   Large residual
X   Unusual X

b. The only slopes that are slightly different based on grade level is the %poverty slope (p-value = 0.017), but with the multiple comparisons, this significance is suspect.
c. The intercepts are not statistically significantly different based on grade (p-value=0.331).

d. The residual plots show no significant violation of assumptions.

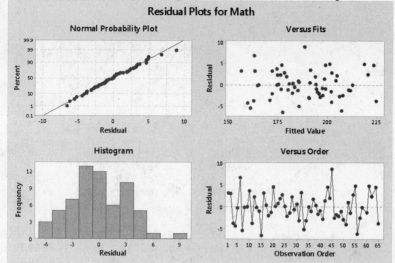

e. By not including a weighting term for the size of the class, we are treating all classes as equal. This may be misleading because a low average score for a small class would carry the same weight as that for a large class while the evidence is not equal.

13.83

a. The regression output is shown below.

Regression Analysis: Math versus Reading, %Minority, %Poverty, Grade

```
Method

Categorical predictor coding   (1, 0)
Rows unused                    3


Analysis of Variance

Source                         DF    Adj SS    Adj MS   F-Value   P-Value
Regression                     29   17601.3   606.942     50.17     0.000
  Reading                       1       6.7     6.693      0.55     0.462
  %Minority                     1       2.6     2.602      0.22     0.646
  %Poverty                      1      10.4    10.365      0.86     0.361
  Grade                         2       2.7     1.331      0.11     0.896
  Reading*Reading               1       6.9     6.925      0.57     0.455
  %Minority*%Minority           1       6.5     6.522      0.54     0.468
  %Poverty*%Poverty             1      35.5    35.503      2.93     0.096
  Reading*%Minority             1       3.6     3.580      0.30     0.590
  Reading*%Poverty              1       8.7     8.725      0.72     0.402
  %Minority*%Poverty            1      14.6    14.625      1.21     0.279
  Reading*Grade                 2       2.8     1.416      0.12     0.890
  %Minority*Grade               2       5.0     2.524      0.21     0.813
  %Poverty*Grade                2      32.9    16.473      1.36     0.270
  Reading*Reading*Grade         2       3.1     1.532      0.13     0.881
  Reading*%Minority*Grade       2       5.2     2.613      0.22     0.807
  Reading*%Poverty*Grade        2      25.2    12.580      1.04     0.365
  %Minority*%Minority*Grade     2      27.8    13.900      1.15     0.329
  %Minority*%Poverty*Grade      2      44.4    22.212      1.84     0.175
  %Poverty*%Poverty*Grade       2      79.2    39.615      3.27     0.050
```

```
Error                          33     399.2   12.097
Total                          62   18000.5
```

Model Summary

```
      S    R-sq  R-sq(adj)  R-sq(pred)
3.47802   97.78%    95.83%      26.84%
```

Coefficients

Term	Coef	SE Coef	T-Value	P-Value	VIF
Constant	2362	2973	0.79	0.433	
Reading	-24.0	32.3	-0.74	0.462	1507979.82
%Minority	-2.52	5.44	-0.46	0.646	93754.71
%Poverty	-5.74	6.20	-0.93	0.361	111736.11
Grade					
4	-1744	5571	-0.31	0.756	36732417.93
5	-1448	3150	-0.46	0.649	10886989.02
Reading*Reading	0.0662	0.0875	0.76	0.455	1603082.40
%Minority*%Minority	0.00298	0.00406	0.73	0.468	517.28
%Poverty*%Poverty	0.00937	0.00547	1.71	0.096	1039.33
Reading*%Minority	0.0160	0.0295	0.54	0.590	81962.19
Reading*%Poverty	0.0284	0.0334	0.85	0.402	96461.21
%Minority*%Poverty	-0.00703	0.00640	-1.10	0.279	1262.42
Reading*Grade					
4	20.9	56.4	0.37	0.714	1.30908E+08
5	15.6	33.7	0.46	0.646	52664114.04
%Minority*Grade					
4	4.89	8.44	0.58	0.566	208966.33
5	0.41	6.69	0.06	0.952	126562.27
%Poverty*Grade					
4	-1.6	16.0	-0.10	0.919	1140805.28
5	10.41	6.90	1.51	0.141	205778.38
Reading*Reading*Grade					
4	-0.061	0.143	-0.42	0.675	29817625.28
5	-0.0425	0.0903	-0.47	0.641	16258119.64
Reading*%Minority*Grade					
4	-0.0274	0.0433	-0.63	0.531	170061.47
5	-0.0077	0.0342	-0.23	0.823	127250.29
Reading*%Poverty*Grade					
4	0.0030	0.0807	0.04	0.970	929331.27
5	-0.0465	0.0354	-1.31	0.198	215918.08
%Minority*%Minority*Grade					
4	-0.00489	0.00878	-0.56	0.581	1217.69
5	-0.00812	0.00536	-1.52	0.139	467.89
%Minority*%Poverty*Grade					
4	0.0047	0.0131	0.36	0.722	3167.94
5	0.0216	0.0113	1.91	0.065	2280.56
%Poverty*%Poverty*Grade					
4	0.0032	0.0101	0.31	0.755	2643.24
5	-0.0244	0.0104	-2.35	0.025	2676.59

```
Regression Equation

Grade
3      Math = 2362 - 24.0 Reading - 2.52 %Minority - 5.74 %Poverty
+ 0.0662 Reading*Reading
             + 0.00298 %Minority*%Minority + 0.00937 %Poverty*%Poverty
+ 0.0160 Reading*
             %Minority + 0.0284 Reading*%Poverty - 0.00703 %Minority*%Poverty

4      Math = 618 - 3.1 Reading + 2.37 %Minority - 7.4 %Poverty
+ 0.006 Reading*Reading
             - 0.00191 %Minority*%Minority + 0.01255 %Poverty*%Poverty -
  0.0114 Reading*
             %Minority + 0.0314 Reading*%Poverty - 0.0023 %Minority*%Poverty

5      Math = 914 - 8.39 Reading - 2.12 %Minority + 4.68 %Poverty
+ 0.0237 Reading*Reading
             - 0.00514 %Minority*%Minority - 0.01499 %Poverty*%Poverty
+ 0.0083 Reading*
             %Minority - 0.0181 Reading*%Poverty + 0.01455 %Minority*%Poverty


Fits and Diagnostics for Unusual Observations

Obs    Math     Fit   Resid   Std Resid
  2   159.60  155.22   4.38      2.58   R
  6   169.80  163.80   6.00      2.13   R
 53   210.20  210.65  -0.45     -1.91         X
 55   205.70  199.86   5.84      2.15   R
 56   201.20  206.71  -5.51     -2.29   R
```

The R^2 and R^2adj are comparable and the second order terms don't appear significant so a first order model is sufficient. The quadratic term for poverty is of moderate significance, but I don't think the value is worth the extra complexity of the model.

b. The prediction is shown below.

Prediction for Math

```
Regression Equation

Math = 2362 - 24.0 Reading - 2.52 %Minority - 5.74 %Poverty + 0.000000 Grade_3 -
  1744 Grade_4
       - 1448 Grade_5 + 0.0662 Reading*Reading + 0.00298 %Minority*%Minority
       + 0.00937 %Poverty*%Poverty + 0.0160 Reading*%Minority
+ 0.0284 Reading*%Poverty
       - 0.00703 %Minority*%Poverty + 0.000000 Reading*Grade_3
+ 20.9 Reading*Grade_4
       + 15.6 Reading*Grade_5 + 0.000000 %Minority*Grade_3 + 4.89 %Minority*Grade_4
       + 0.41 %Minority*Grade_5 + 0.000000 %Poverty*Grade_3 - 1.6 %Poverty*Grade_4
       + 10.41 %Poverty*Grade_5 + 0.000000 Reading*Reading*Grade_3
       - 0.061 Reading*Reading*Grade_4 - 0.0425 Reading*Reading*Grade_5
+ 0.000000 Reading*
       %Minority*Grade_3 - 0.0274 Reading*%Minority*Grade_4 - 0.0077 Reading*
       %Minority*Grade_5 + 0.000000 Reading*%Poverty*Grade_3 + 0.0030 Reading*
       %Poverty*Grade_4 - 0.0465 Reading*%Poverty*Grade_5 + 0.000000 %Minority*
       %Minority*Grade_3 - 0.00489 %Minority*%Minority*Grade_4 - 0.00812 %Minority*
       %Minority*Grade_5 + 0.000000 %Minority*%Poverty*Grade_3 + 0.0047 %Minority*
       %Poverty*Grade_4 + 0.0216 %Minority*%Poverty*Grade_5 + 0.000000 %Poverty*
       %Poverty*Grade_3 + 0.0032 %Poverty*%Poverty*Grade_4 -
  0.0244 %Poverty*%Poverty*Grade_5
```

```
Variable    Setting
Reading       170
%Minority      40
%Poverty       30
Grade           3
```

```
  Fit    SE Fit       95% CI              95% PI
180.814  6.19899  (168.202, 193.426)  (166.353, 195.276)   XX
```

XX denotes an extremely unusual point relative to predictor levels used to fit the model.

The fit is 180.814 and the 95% CI is (168.202, 193.426).

c. While the fits differ by approximately 10 points, the confidence intervals overlap so there is no significant difference.

Chapter 14

Analysis of Variance for Completely Randomized Designs

14.1   Assign the numbers 1, 2, …, 20 to the 20 students. Use a random number table or computer software to obtain a random permutation of the 20 numbers. The first 5 numbers in the list are assigned to the control group, the next 5 the piano lessons, and so forth.

14.3
   a. The ANOVA table is given here:

Source	DF	SS	MS	F	p-value
Instruction	3	727.24	242.41	24.44	0.000
Error	96	952.30	9.92		
Total	99	1679.54			

   b. The p-value for Instruction is 0.000 < 0.05 which implies there is a difference in mean effectiveness of the methods of instruction.

   c. Using Tukey's W procedure, the Minitab output is shown here:

   One-way ANOVA: Control, Piano, Computer, Instruct

   ```
   Source  DF      SS      MS      F       P
   Factor   3  727.24  242.41  24.44   0.000
   Error   96  952.30    9.92
   Total   99 1679.54

   S = 3.150   R-Sq = 43.30%   R-Sq(adj) = 41.53%
   ```

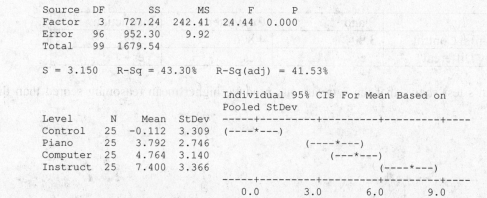

   ```
   Pooled StDev = 3.150


   Tukey 95% Simultaneous Confidence Intervals
   All Pairwise Comparisons

   Individual confidence level = 98.97%

   Control subtracted from:
   ```

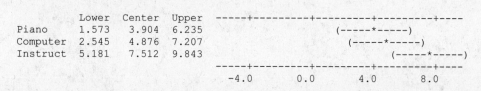

187

```
Piano subtracted from:

            Lower   Center   Upper    -----+---------+---------+---------+----
Computer   -1.359   0.972   3.303              (----*-----)
Instruct    1.277   3.608   5.939                     (-----*-----)
                                      -----+---------+---------+---------+----
                                         -4.0      0.0       4.0       8.0


Computer subtracted from:

            Lower   Center   Upper    -----+---------+---------+---------+----
Instruct    0.305   2.636   4.967                    (-----*----)
                                      -----+---------+---------+---------+----
                                         -4.0      0.0       4.0       8.0
```

From the output, there is a significant difference between the Control and all other methods, between Piano and Instructor and between Computer and Instructor. This is because the confidence intervals do not include 0.

d. The Dunnett's one-sided multiple comparison procedure is employed to compare the treatments to the control with a family-wise error rate of 0.05.

$$D = d_{0.05}(3,96)\sqrt{\frac{2(9.92)}{25}} = (2.09)(.89) = 1.86$$

Treatment	Control	Piano	Computer	Instruction
Mean	-0.112	3.792	4.764	7.400

Treatment	Piano	Computer	Instruction
Difference from Control	3.904	4.876	7.512
Significantly Different?	Yes	Yes	Yes

The Dunnett's test shows that the 3 treatments produce higher mean reasoning scored than the control.

14.5

a. H_0: All incentive plan yield equal productivity vs. H_a: at least one incentive plan yields differing productivity

b. The p-value on the AOV table is $0.118 > 0.05$. This means there is not significant evidence that the 4 incentive plans yield different productivity.

c. The Minitab output is given below

```
Pooled StDev = 197.7

Tukey 95% Simultaneous Confidence Intervals
All Pairwise Comparisons

Individual confidence level = 98.97%

Plan  A subtracted from:

            Lower   Center   Upper
Plan  B    -121.3     43.2   207.7
Plan  C     -50.2    114.3   278.7
Plan  D     -28.8    135.7   300.2
```
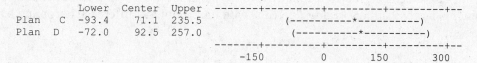

```
Plan  B subtracted from:

            Lower   Center   Upper
Plan  C     -93.4    71.1   235.5
Plan  D     -72.0    92.5   257.0
```
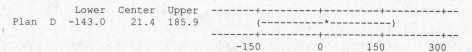

```
Plan  C subtracted from:

            Lower   Center   Upper
Plan  D    -143.0    21.4   185.9
```

Because all C.I.'s include 0, there is not a significant difference in any pair of incentive plans.

14.7

a. The p-value for the Kruskal-Wallis test is 0.025 which suggests there is sufficient evidence to reject the hypothesis of equality of incentive plans.

b. Individual student solution

c. The residuals appear to have a right skewed distribution which reduces the ability of the ANOVA F test to detect differences in the treatment means. The Kruskal-Wallis test, by analyzing the ranks of the data, is less affected by the large residuals.

14.9

a. A profile plot of the data is given here

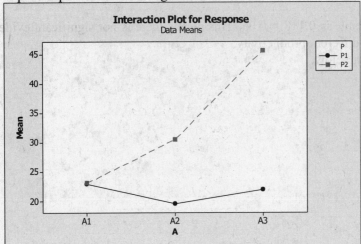

b. The p-value for the interaction team is 0.013. There is significant evidence of an interaction between the factors Age and Product Type. Thus, the size of the difference between mean attention span of children viewing breakfast cereal ads and viewing video game ads would be different for the three age groups. From the profile plots, the estimated mean attention span for video games is larger than for breakfast cereals, with the size of the difference becoming larger as age increases.

14.11

a. Completely randomized design with a 3x3x2 factorial treatment structure and 3 reps.
b. A model for this experiment is given here:
$y_{ijkm} = \mu + \alpha_i + \beta_j + (\alpha\beta)_{ij} + \gamma_k + (\alpha\gamma)_{ik} + (\beta\gamma)_{jk} + (\alpha\beta\gamma)_{ijk} + \varepsilon_{ijkm}$; with
$i = 1,2,3;\ j = 1,2,3;\ k = 1,2;\ m = 1,2,3$
where y_{ijkm} is the sensory rating of the ice cream made from the ith milk fat level, j^{th} amount of air, and the kth sweetener.
α_i is the effect of the ith milk fat level on the sensory rating
β_j is the effect of the jth amount of air on the sensory rating
γ_k is the effect of the kth amount of sweetener on the sensory rating
$(\alpha\beta)_{ij}$ is the interaction effect of the ith milk fat level and j^{th} amount of air on sensory rating
$(\alpha\gamma)_{ik}$ is the interaction effect of the ith milk fat level and k^{th} amount of sweetener on sensory rating
$(\beta\gamma)_{jk}$ is the interaction effect of the jth amount of air and k^{th} amount of sweetener on sensory rating
$(\alpha\beta\gamma)_{ijk}$ is the interaction effect of the ith milk fat level, j^{th} amount of air, and k^{th} amount of sweetener on sensory rating

c. The profile plots are given here (Sweetener 12% on the left and 16% on the right):

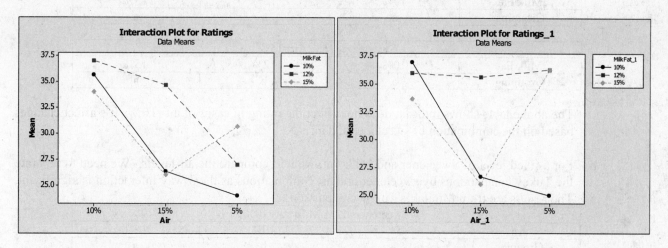

d. The interactions look significant on each plot and yet different for the two levels of sweetener—noticeably for 12% milk fat level. This leads to the conclusion that a three-way interaction exists.

14.13
a. The residuals in the normal probability plot appear to fall very close to a straight line and hence we can conclude there is not significant evidence that the residuals have a non-normal distribution.
b. The plot of the residuals vs. Fitted Value appears to have a consistent width across the fitted values. The condition of constant variance does not appear to be violated.
c. We could discuss with the experimenter the manner in which the experiment was conducted and the data were collected, making sure to inquire about any potential temporal or spatial correlation.

14.15
a. For a fixed level of Sweetener and Air, we will compare the milk fat levels. We need to separate the Tukey comparisons by sweetener and air combinations as the 3 way interaction is significant. The results are from Minitab's Tukey comparison.

Sweetener=12% Air=5%			
Milk Fat	10%	12%	15%
Grouping	a	ab	b

Sweetener=12% Air=10%			
Milk Fat	10%	12%	15%
Grouping	a	a	a

Sweetener=12% Air=15%			
Milk Fat	10%	12%	15%
Grouping	a	b	a

Sweetener=16% Air=5%			
Milk Fat	10%	12%	15%
Grouping	a	b	b

Sweetener=16% Air=10%			
Milk Fat	10%	12%	15%
Grouping	a	a	a

Sweetener=16% Air=15%			
Milk Fat	10%	12%	15%
Grouping	a	b	a

The above table shows milk fat does not affect the rating in cases of air=10%. It's affect changes based on the combination of sweetener and air.

b. For a fixed level of Sweetener and Milk Fat, we will compare the air levels. We need to separate the Tukey comparisons by sweetener and air combinations as the 3 way interaction is significant. The results are from Minitab's Tukey comparison.

Sweetener=12% Milk Fat=10%			
Air	5%	10%	15%
Grouping	a	b	a

Sweetener=12% Milk Fat=12%			
Air	5%	10%	15%
Grouping	a	a	b

Sweetener=12% Milk Fat=15%			
Air	5%	10%	15%
Grouping	a	b	a

Sweetener=16% Milk Fat=10%			
Air	5%	10%	15%
Grouping	a	b	a

Sweetener=16% Milk Fat=12%			
Air	5%	10%	15%
Grouping	a	a	a

Sweetener=16% Milk Fat=15%			
Air	5%	10%	15%
Grouping	a	a	b

The above table shows the effect of air percentage changes based on the combination of sweetener and milk fat.

c. The combinations producing the highest mean sensory rating (37) are milk=12%, air=10%, and sweetener=12% or milk=10%, air=10%, and sweetener=16%.

14.17 The necessary parameters are $t = 2^3 = 8$, $D = 20$, $\alpha = 0.05$, $\sigma = 9$

$$\phi = \sqrt{\frac{r(20)^2}{(2)(8)(9)^2}} = 0.6415\sqrt{r}$$

Determine r so that power is .80. Select values for r, compute $v_1 = t - 1 = 8 - 1 = 7$
$v_2 = t(r - 1) = 8(r - 1)$, and $\phi = 0.5556\sqrt{r}$, then use Table 14 with $\alpha = 0.05$ and $t = 8$ to determine power:

r	v_2	ϕ	Power
5	32	1.24	.63
6	40	1.36	.68
7	48	1.46	.81

Thus, it would take 7 reps to obtain a power of at least .80.

14.19

The necessary parameters are $t = 18$, $D = 5$, $\alpha = 0.05$, $\sigma = 1.81$

$$\phi = \sqrt{\frac{r(5)^2}{(2)(18)(1.81)^2}} = 0.4604\sqrt{r}$$

Determine r so that power is .90. Select values for r, compute $v_1 = t - 1 = 18 - 1 = 17$ $v_2 = t(r-1) = 18(r-1)$, and $\phi = 0.4604\sqrt{r}$, then using computer software with $\alpha = 0.05$ and $t = 18$ to determine power:

r	v_2	ϕ	Power
7	108	1.22	.88
8	126	1.30	.94

Thus, it would take about r = 8 reps to obtain a power of at least .90.

14.21

a. The test for interaction has F = 11.34 with df = 9,16 which yields p-value < 0.0001. This implies there is significant evidence of an interaction between Cu Rate and Mn Rate on Soybean yield.
b. Mn = 110
c. Cu = 7
d. (Cu,Mn) = (7,110)

14.23

a. The profile plot is given here:

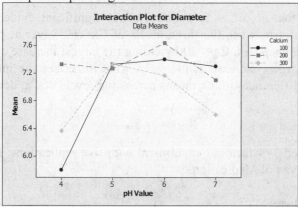

There appears to be an interaction between Ca Rate and pH with respect to the increase in trunk diameters. At low pH values, a 300 level of Ca yields the largest increase; whereas, at high pH values, a 100 level of Ca yields the largest increase in trunk diameter.

b. $y_{ijk} = \mu + \tau_i + \beta_j + (\tau\beta)_{ij} + \varepsilon_{ijk}$; $i = 1,2,3,4$; $j = 1,2,3$; $k = 1,2,3$;
Where y_{ijk} is the increase in trunk diameter of the kth tree in soil having the ith pH level using the jth Ca Rate;
τ_i is the effect of the ith pH level on diameter increase
β_j is the effect of the jth Ca Rate on diameter increase
$(\tau\beta)_{ij}$ is the interaction effect of the ith pH level and j^{th} Ca Rate on diameter increase

c. The ANOVA table is given here:

Source	DF	SS	MS	F	p-value
pH	3	4.461	1.487	21.94	0.0001
Ca	2	1.467	0.734	10.82	0.0004
Interaction	6	3.255	0.543	8.00	0.0001
Error	24	1.627	0.0678		
Total	35	10.810			

The design is a completely randomized 3x4 factorial design with 3 replicates.

14.25

a. Using Tukey's W procedure with $\alpha = 0.05, s_\epsilon^2 = MSE = 0.0678$, $q_\alpha(t, df_{error}) = q_{.05}(3, 24) = 3.53 \Rightarrow$

$$W = (3.53)\sqrt{\frac{0.0678}{3}} = 0.53 \Rightarrow$$

		Ca Rate		
pH		100	200	300
4	Mean	5.80	7.33	6.37
	Grouping	a	c	b
5	Mean	7.33	7.27	7.33
	Grouping	a	a	a
6	Mean	7.40	7.63	7.17
	Grouping	a	a	a
7	Mean	7.30	7.10	6.60
	Grouping	b	ab	a

b. From the above table, we observe that at pH = 5, 6 there is not significant evidence of a difference in mean increase in diameter between the three levels of Ca. However at pH = 4, 7 there is significant evidence of a difference with Ca = 300 yielding the largest increase at pH = 4 and Ca = 100 or 200 yielding the largest increase at pH = 7. This illustrates the interaction between Ca and pH, i.e., the size of differences in the means across the levels of Ca, depends on the level of pH.

14.27

a. The design is a completely randomized 3×9 factorial experiment with five replications; Factor A is Level of Severity and Factor B is Type of Medication.

b. A model for this experiment is given here:
$y_{ijk} = \mu + \tau_i + \beta_j + (\tau\beta)_{ij} + \varepsilon_{ijk}$; $i = 1,2,3$; $j = 1, ..., 9$; $k = 1,2,3,4,5$;
Where y_{ijk} is the temperature of the kth patient having the ith severity level using the jth medication;
τ_i is the effect of the ith severity level on temperature
β_j is the effect of the jth medication on temperature
$(\tau\beta)_{ij}$ is the interaction effect of the ith severity level and j^{th} medication on temperature

14.29
a. This a completely randomized design with a single factor, Weight Classification, having four levels. There are seven replication of the experiment at each weight classification. A model for this experiment is:
$y_{ij} = \mu + \tau_i + \varepsilon_{ij}$; $i = 1,2,3,4$; $k = 1,\ldots,7$
where y_{ij} is the temperature of the jth patient having the ith weight classification
τ_i is the effect of the ith weight classification on fatigue time

b. The complete AOV table is given here:

Source	DF	SS	MS	F	p-value
Wt. Class.	3	892.7	297.6	11.06	0.0001
Error	24	646.0	26.9		
Total	27	1538.7			

There is significant evidence, p-value < 0.0001, that the average fatigue time is different for the four weight classifications.

14.31
a. The design is a completely randomized 5x5 factorial experiment with 2 replications; Factor A is Exterior Temperature and Factor B is Pane Design.
A model for this experiment is given here:
$y_{ijk} = \mu + \tau_i + \beta_j + (\tau\beta)_{ij} + \varepsilon_{ijk}$; $i = 1,2,3$; $j = 1,2$; $k = 1,\ldots,10$;
where y_{ijk} is the heat loss of the kth pane having the ith temperature level and j^{th} pane design.
τ_i is the effect of the ith temperature on heat loss
β_j is the effect of the jth pane design on heat loss
$(\tau\beta)_{ij}$ is the interaction effect of the ith temperature and j^{th} pane design on heat loss

b. The test for an interaction between exterior temperature in pane design yields p-value = 0.0073 which would indicate significant evidence that an interaction exists. Therefore, the difference in mean heat loss between the five pane designs varies depending on the exterior temperature. The test of the main effect of pane design is not informative due to the significant interaction.

c. No, because of the significant interaction between pane design and exterior temperature. A profile plot is given here along with the table of treatment means:

	Pane Design				
Temperature	A	B	C	D	E
0	10.50	10.50	11.45	11.60	9.60
20	9.50	9.50	10.45	10.60	9.55
40	9.45	9.50	10.40	10.45	9.45
60	8.10	8.45	8.55	8.45	9.55
80	7.50	7.50	8.45	8.60	9.55

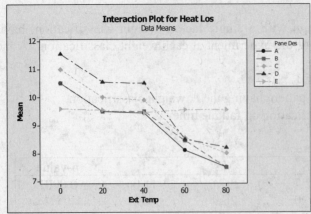

d. Using Tukey's W procedure with $\alpha = 0.05$, $s_\epsilon^2 = MSE = 0.2312$, $q_\alpha(t, df_{error}) = q_{.05}(5,25) = 4.16 \Rightarrow W = (4.16)\sqrt{\frac{0.2312}{2}} = 1.41 \Rightarrow$

	Pane Design				
Temperature	A	B	C	D	E
Temp=0 Mean	10.50	10.50	11.45	11.60	9.60
Grouping	ab	ab	b	b	a
Temp=20 Mean	9.50	9.50	10.45	10.60	9.55
Grouping	a	a	a	a	a
Temp=40 Mean	9.45	9.50	10.40	10.45	9.45
Grouping	a	a	a	a	a
Temp=60 Mean	8.10	8.45	8.55	8.45	9.55
Grouping	a	a	a	a	b
Temp=80 Mean	7.50	7.50	8.45	8.60	9.55
Grouping	a	a	a	a	b

From the above table, we observe that at the exterior temperatures of 20°F and 40°F there is not significant evidence of a difference in mean heat loss between the five pane designs. However, at the exterior temperatures of 60°F and 80°F, pane design E has a significantly higher mean heat loss than the other four designs. At the exterior temperature of 0°F, there are two groups of pane designs relative to their mean heat loss. This illustrates the interaction between exterior temperature and pane design, i.e., the size of differences in mean heat loss between the five pane designs depends on the external temperature.

14.33
a. The experiment is run as three reps of a completely randomized design with a 2x4 factorial treatment structure. A model for this experiment is given here:
$y_{ijk} = \mu + \tau_i + \beta_j + (\tau\beta)_{ij} + \varepsilon_{ijk}$; $i = 1,2,3,4$; $j = 1,2$; $k = 1,2,3$;
where y_{ijk} is the amount of active ingredient (or pH) of the kth vial having the ith storage time and j^{th} laboratory.
τ_i is the effect of the ith storage time on amount of active ingredient (or pH)
β_j is the effect of the jth laboratory on amount of active ingredient (or pH)
$(\tau\beta)_{ij}$ is the interaction effect of the ith storage time and j^{th} laboratory on amount of active ingredient (or pH)

b. The complete AOV table is given here:

Source	DF	SS	MS	F	p-value
Storage Time	3	SSA	SSA/3	MSA/MSE	
Laboratory	1	SSB	SSB/1	MSB/MSE	
Interaction	3	SSAB	SSAB/3	MSAB/MSE	
Error	16	SSE	SSE/16		
Total	23	SSTot			

14.35

a. The test for an interaction yields p-value = 0.0255. There is significant evidence that an interaction exists between Ratio and Supply in regards to the mean Profit. The following profile plot displays the interaction:

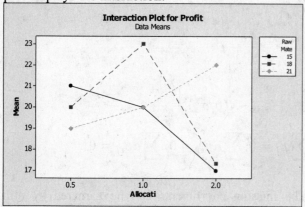

b. Because of the significant interaction between the factors Ratio and Supply, it is not possible to consider the factors separately. Therefore, the nine treatments consisting of nine different combinations of Ratio and Supply will be examined using LSD-procedure with $\alpha = 0.05, s_\epsilon^2 = MSE = 4.592593, t_{0.025,18} = 2.101 \Rightarrow$

$$LSD = (2.101)\sqrt{\frac{(2)(4.592593)}{3}} = 3.68 \Rightarrow$$

	Ratio-Supply								
	2:15	2:18	.5:21	.5:18	1:15	1:21	.5:15	2:21	1:18
Mean	17	17.33	19	20	20	20	21	22	23
Grouping	a	ab	abc	abcd	abcd	abcd	bcd	cd	d

The combinations of Ratio and Supply yielding the highest mean profits are
(Ratio=.5, Supply=15), (Ratio=.5, Supply=18), (Ratio=1, Supply=15),
(Ratio=1, Supply=18), (Ratio=1, Supply=21), and (Ratio=2, Supply=21)
These six combinations do not have significantly different mean profits.

14.37

a. The computational formulas are not applicable in this case as the design is not balanced.
b. The residual plots are shown below. There are no major violations to the assumptions.

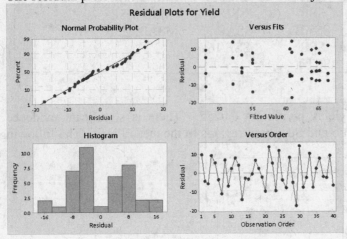

c. $y_{ijk} = \mu + \alpha_i + \beta_j + \alpha\beta_{ij} + \varepsilon_{ijk}$; $\varepsilon_{ijk} \sim N(0, \sigma^2)$;

$$\sum \alpha_i = 0; \sum \beta_j = 0; \sum_i \alpha\beta_{ij} = 0; \sum_j \alpha\beta_{ij} = 0$$

The following are the estimates. The missing coefficients can be garnered by utilizing the conditions above.

```
Coefficients

Term                         Coef   SE Coef  T-Value  P-Value   VIF
Constant                    59.75      1.55    38.50    0.000
Growth Stage
  Early                     -3.27      2.22    -1.48    0.151  1.41
  Late                       3.90      2.15     1.82    0.080  1.37
Percent Shading
  0.00%                     -1.45      2.57    -0.56    0.577  1.53
  25.00%                    -0.68      2.80    -0.24    0.809  1.65
  50.00%                     0.10      2.73     0.04    0.972  1.64
Growth Stage*Percent Shading
  Early  0.00%               5.64      3.76     1.50    0.145  2.17
  Early 25.00%              -7.97      3.76    -2.12    0.043  2.03
  Early 50.00%               5.87      4.18     1.40    0.171  2.47
  Late  0.00%               -1.28      3.55    -0.36    0.721  1.94
  Late 25.00%                1.66      3.88     0.43    0.672  1.83
  Late 50.00%                1.38      3.66     0.38    0.710  2.07
```

14.39

a. The ANOVA table is shown below.

```
Analysis of Variance

Source              DF   Adj SS   Adj MS  F-Value  P-Value
  City               2   5720.7  2860.34   149.13    0.000
  Sludge Rate        2   1945.4   972.72    50.71    0.000
  City*Sludge Rate   4   1809.4   452.35    23.58    0.000
Error               27    517.9    19.18
Total               35   9993.4
```

There is significant evidence of an interaction between the city and sludge rate on the zinc content. The main effects are also significant but interpretation of these tests is suspect due to the presence of an interaction.

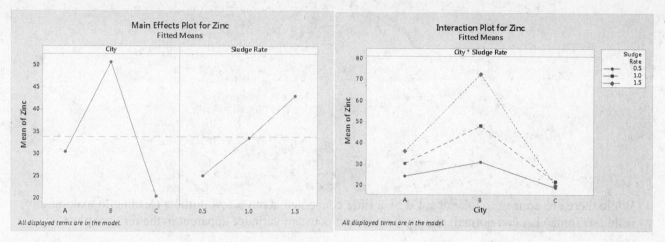

b. Linear trend analysis

Sludge	Mean	n
0.5	24.94	12
1.0	33.52	12
1.5	42.94	12

$$\hat{l} = (-0.5)(24.94) + (0.5)(42.94) = 9$$

$$SSC = \frac{9^2}{\frac{0.5^2}{12} + \frac{0.5^2}{12}} = 1944$$

$$F = \frac{1944}{19.81} = 98.1323$$

This F is larger than any reasonable critical value which implies there is a significant linear trend.

c. Based on the Tukey tests, all three pairs are significantly different but we must be careful about making claims on the sludge rate without considering the city as there is an interaction present.

Tukey Pairwise Comparisons: Response = Zinc, Term = Sludge Rate

```
Grouping Information Using the Tukey Method and 95% Confidence

Sludge
Rate    N     Mean  Grouping
1.5    12  42.9417  A
1.0    12  33.5167       B
0.5    12  24.9417            C

Means that do not share a letter are significantly different.
```

14.40.

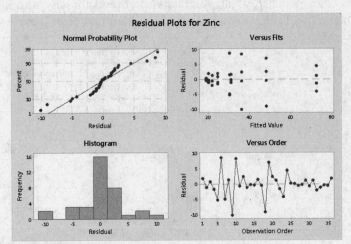

While there are no major violations, I'd be a little concerned about a few outliers (leading to making the residuals somewhat non-normal) and a potential non-constant variance apparent in the residual plots.

Chapter 15

Analysis of Variance for Blocked Designs

15.1
  a. Assign the numbers 1, 2, 3, 4, 5 to the plots on each farm. Separately for each farm, obtain a random ordering of the numbers 1, 2, 3, 4, 5. The plot with the first number on the list is assigned to Surface, second to Trickle, etc. For example, Farm 1—2, 3, 1, 5, 4; Farm 2—3, 5, 2, 4, 1...

Farm	Surface	Trickle	Center	Lateral	Sub
1	P2	P3	P1	P5	P4
2	P3	P5	P2	P4	P1
...	...	...	...	...	...
10	P5	P1	P4	P2	P3

  b. $5! = 120$
  c. $(120)^{10} = 6.195 \times 10^{20}$

15.3
  a. $t = 5, b = 10 \Rightarrow RE(RCB, CR) = \frac{(10-1)(7369)+(10)(5-1)(1725)}{((10)(5)-1)(1725)} = 1.6$
  b. It would take 1.6 times as many observations (approximately 16) per treatment in a completely randomized design to achieve the same level of precision in estimating the treatment means as was accomplished in the randomized complete block design.

15.5
  a. **Paired T-Test and CI: Device1, Device2**
```
Paired T for Device1 - Device2

              N    Mean   StDev  SE Mean
Device1      24   3.541   0.605    0.124
Device2      24   2.450   0.565    0.115
Difference   24   1.091   0.926    0.189

95% CI for mean difference: (0.700, 1.482)
T-Test of mean difference = 0 (vs not = 0): T-Value = 5.77   P-Value = 0.000
```
The p-value for the test of differing means is 0.000 which yields the same conclusion in Exercise 15.4.

  b. The t statistic given in part A is 5.77 and the F statistic from Exercise 15.4 is $33.2929 = 5.77^2$.
  $$t^2_{.025,23} = 2.06866^2 = 4.27934 = F_{.05,1,23}$$

15.7 The model conditions appear to be satisfied:
  --The normal probability plots and box plots of the residuals do not indicate nonnormality.
  --Plot of residuals vs. estimated mean does not indicate nonconstant variance
  --Interaction plot indicates a potential interaction between subjects and type of music, but the indications are fairly weak.

15.9
a. Divide the farm into 4 rows and 4 columns. Randomly assign the numbers 1, 2, 3, 4 to the 4 treatments. Randomly assign the 4 treatments to the 16 locations on the farm such that each of the four treatments appears once and only once in each column and each row (Latin Square design).

b. $RE(LS, CR) = \frac{MSR+MSC+(t-1)MSE}{(t+1)MSE} = \frac{0.000283+0.004117+(4-1)(0.000125)}{(4+1)(0.000125)} = 7.64 \Rightarrow$

It would take 7.64 times as many observations (approximately 30) per treatment in a completely randomized design to achieve the same level of precision in estimating the treatment means as was accomplished in the Latin square design.

c. It appears that the row block is not helpful in reducing the sum of squares after the column effect is accounted for. The p-value for Row is $0.181 > 0.05$.

15.11
a. Similar results
b. --The boxplot and normal probability plot do not indicate deviation from a normal distribution for the residuals.
--The plot of Residuals vs. Pred do not indicate a deviation from the constant variance assumption
--Based on these plots, there is no indication of deviation from the model conditions.

15.13
a. There are 24 treatments consisting of the 2 Chemicals, 3 levels of dose, and 4 environmental conditions. For each colony, select 24 'groups' of 100 ants and label them 1-24. Assign each number to a specific treatment combination.

b. The table is given here:

Source	DF
Colonies (Blocks)	4
Chemical	1
Dose	2
C x D	2
Environment	3
C x E	3
D x E	6
C x D x E	6
Error	93
Total	119

15.15
a. Randomly assign the numbers 1, 2, ..., 6 to the 6 road segments within each of the 5 types of roadways. Separately for each type of roadway, randomly assign one of the 6 road segments to one of the 6 treatments. For example,

	Sodium Chloride		Calcium Chloride		Sand	
Roadway	L	H	L	H	L	H
1	S2	S5	S4	S3	S6	S1
2	S3	S4	S5	S2	S1	S6
3	S6	S3	S2	S1	S5	S4
4	S4	S2	S1	S3	S6	S5
5	S1	S3	S5	S6	S4	S2

b. $t = 6, b = 5 \Rightarrow RE(RCB, CR) = \frac{(5-1)(243.28)+(5)(6-1)(3.76)}{((5)(6)-1)(3.76)} = 9.79$

It would take 9.79 times as many observations (approximately 50) per treatment in a completely randomized design to achieve the same level of precision in estimating the treatment means as was accomplished in the randomized complete block design. Since RE was much larger than 1, we would conclude the blocking was effective.

c. Based on the RE, blocking by traffic volume was extremely efficient so the practice should be continued in future studies.

15.17
a. There are 12 treatments consisting of the 4 levels of planting density and 3 levels of tomato variety. Randomly assign the numbers 1, 2,…, 12 to the 12 plots in each field. These 12 treatments will be randomly assigned to 12 plots in each of the three fields as seen in the following diagram:

	Field 1			Field 2			Field 3		
	Variety			Variety			Variety		
Planting density	Celebrity	Sunbeam	Trust	Celebrity	Sunbeam	Trust	Celebrity	Sunbeam	Trust
5k	P5	P6	P9	P10	P4	P7	P1	P4	P7
20k	P8	P1	P12	P6	P8	P12	P5	P3	P6
35k	P11	P2	P4	P11	P3	P5	P12	P8	P10
50k	P3	P7	P10	P1	P9	P2	P2	P9	P11

b. $t = 12, b = 3 \Rightarrow RE(RCB, CR) = \frac{(3-1)(265.71)+(3)(12-1)(4.76)}{((3)(12)-1)(4.76)} = 4.13$

It would take 4.13 times as many observations (approximately 12) per treatment in a completely randomized design to achieve the same level of precision in estimating the treatment means as was accomplished in the randomized complete block design. Since RE was much larger than 1, we would conclude the blocking was effective.

c. It was shown in part b that the blocking was effective in removing variability due to field. If the same tomatoes were planted in each field, the effect of the field and variety would be confounded. This means that any differences that are observed in the treatment means could be due to variety or due to field. There would be no way to separate out the effects of variety and field.

15.19
a. There are three conditions to be satisfied to run Friedman's test
1. The experimental design is a randomized block design, with the t treatments randomly assigned to exactly one experimental unit per block yielding $N = tb$ responses.
2. The N responses y_{ij} are mutually independent.
3. The N responses are related by the model $y_{ij} = \theta + \tau_i + \beta_j + \epsilon_{ij}$ where θ is the overall median, τ_i is the effect due to the ith treatment, β_j is the effect due to the jth block, and the N ε_{ij}'s are a random sample from a continuous distribution with a median equal to 0.

b. The Minitab output is shown below:
Friedman Test: Weight versus Method blocked by Farm

```
S = 37.04   DF = 4   P = 0.000

                        Est      Sum of
Method           N    Median     Ranks
CenterPoint      10   469.4      32.0
Lateral          10   500.6      38.0
SubIrrigation    10   396.0      17.0
Surface          10   622.9      50.0
Trickle          10   360.9      13.0

Grand median = 469.9
```

The p-value for equality of irrigation methods is 0.000 which means there is significant evidence that the different irrigation methods yield differing blueberry yields.

c. The results are the same as the results from the AOV F-test from 15.2.
d. These results are expected to match up with the F test because the conditions for the F test are satisfied and hence both the F test and Friedman's test are valid.

15.21
a. The Minitab output is shown below:

Friedman Test: Yield versus Treatment blocked by Field

```
S = 31.62   DF = 11   P = 0.001
S = 31.69   DF = 11   P = 0.001 (adjusted for ties)

                        Est      Sum of
Treatment        N    Median     Ranks
Celebrity20K     3    44.91      14.0
Celebrity35K     3    47.61      16.0
Celebrity50K     3    43.47       9.0
Celebrity5K      3    36.92       4.5
Sunbeam  20K     3    48.01      18.0
Sunbeam  35K     3    52.23      23.5
Sunbeam  50K     3    48.45      18.5
Sunbeam  5K      3    36.95       4.5
Trust    20K     3    63.91      32.0
Trust    35K     3    70.44      36.0
Trust    50K     3    63.15      30.0
Trust    5K      3    56.48      28.0

Grand median = 51.05
```

The p-value for equality of treatment combinations is 0.001 which means there is significant evidence that the different variety/density combinations yield differing tomato amounts.

b. The results are the same as the results from the AOV F-test from 15.16.

15.23
- a. $y_{ij} = \mu + \alpha_i + \beta_j + \epsilon_{ij}$; $i = 1,2,3,4$; $j = 1,2,3,4,5$
  y_{ij} is the measurement by the jth investigator of the ith mixture
  α_i is the ith mixture effect
  β_j is the jth investigator effect
- b. The estimates of the model parameters are not useful in assessing the effects of the treatments.
- c. $F = 1264.73$ with p-value $< 0.0001 \Rightarrow$
  Reject $H_0: \mu_1 = \mu_2 = \mu_3 = \mu_4$ and conclude there is significant evidence of a difference in the mean propellant thrust for the four mixtures.
- d. Mixture 2 with the highest mean response would appear to be the best mixture. A multiple comparison procedure could be used to confirm that the other three mixtures have significantly lower means.
- e. $RE(RCB, CR) = \frac{(b-1)MSB + b(t-1)MSE}{(bt-1)MSE} = \frac{(5-1)(113.12) + (5)(4-1)(68.86)}{((5)(4)-1)(68.86)} = 1.14 \Rightarrow$
  It would take 1.14 times as many observations (approximately 6) per treatment in a completely randomized design to achieve the same level of precision in estimating the treatment means as was accomplished in the randomized complete block design.

15.25 The model conditions appear to be satisfied:
--The normal probability plots and box plots of the residuals do not indicate nonnormality.
--Plot of residuals versus estimated mean does not indicate nonconstant variance
--Interaction plot indicates that the additive model (no interaction between attitude and type of workshop) is valid.

15.27
- a. Because all the students were in the same grade, this is a completely randomized design with a 2x3 factorial treatment structure. Factor A-Sex and Factor B-Level of Abuse (3 levels). There are 30 reps of the complete experiment.
- b. The grade level factor is considered as a third factor since age, as reflected by grade level, may interact with sex, because girls tend to mature more rapidly than boys. Thus, the design would be a completely randomized design with a 2x3x3 factorial treatment design: Factor A-Sex, Factor B-Level of Abuse (3 levels), Factor C-Grade Level (3 levels). There are 10 reps of the complete experiment.

15.29
- a. A profile plot of the data is given here:

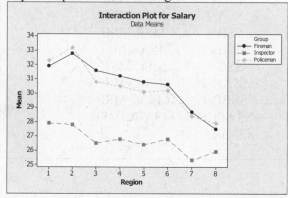

Based on the profile plot, the additive model appears to be appropriate because the three lines are relatively parallel. Note further that the plotted points are means of a single observation and hence there may be considerable variability in their estimation of the population means μ_{ij}. Thus, exact parallelism is not required in the profile plots to ensure the validity of the additive model.

It would not be possible to test for an interaction between Region and Job Type because there is only one observation per Region-Job Type combination.

b. $RE(RCB, CR) = \frac{(b-1)MSB + b(t-1)MSE}{(bt-1)MSE} = \frac{(8-1)(6.089) + (8)(3-1)(0.422)}{((8)(3)-1)(0.422)} = 5.09 \Rightarrow$

It would take 5.09 times as many observations (approximately 41) per treatment in a completely randomized design to achieve the same level of precision in estimating the treatment means as was accomplished in the randomized complete block design.

c. Other possible important factors may be average salaries of all government employees in the region, education requirements for position, etc.

15.31

a. A randomized complete block design with a 3x2x2 factorial treatment structure. The blocks are the six panels, Factor A is Sweetness, Factor B is Caloric Content, and Factor C is Color. There is one replication of the complete experiment.

b. A model for this experiment is given here:

$y_{ijkm} = \mu + v_i + \alpha_j + \beta_k + \alpha\beta_{jk} + \gamma_m + \alpha\gamma_{jm} + \beta\gamma_{km} + \alpha\beta\gamma_{jkm} + \epsilon_{ijkm}$;

with $i = 1, \ldots, 6; j = 1,2,3; k = 1,2; m = 1,2$;

where y_{ijkm} is the rating of the ith panel of a drink formulated with the jth sweetness level, k^{th} caloric level, and m^{th} color

v_i is the effect of the ith panel on rating

α_j is the effect of the jth sweetness level on rating

β_k is the effect of the kth caloric level on rating

$\alpha\beta_{jk}$ is the interaction effect of the jth sweetness level and k^{th} caloric level on rating

γ_m is the effect of the mth color on rating

$\alpha\gamma_{jm}$ is the interaction effect of the jth sweetness level and m^{th} color on rating

$\beta\gamma_{km}$ is the interaction effect of the kth caloric level and m^{th} color on rating

$\alpha\beta\gamma_{jkm}$ is the interaction effect of the jth sweetness level, k^{th} caloric level, and m^{th} color on rating

c. The complete AOV table is given here:

Source	DF	SS	MS	F	p-value
Panels	5	SSP	SSP/5		
Treatments	11	SST	SST/11	MST/MSE	
Sweetness (A)	2	SSA	SSA/2	MSA/MSE	
Caloric (B)	1	SSB	SSB/1	MSB/MSE	
AB	2	SSAB	SSAB/2	MSAB/MSE	
Color (C)	1	SSC	SSC/1	MSC/MSE	
AC	2	SSAC	SSAC/2	MSAC/MSE	
BC	1	SSBC	SSBC/1	MSBC/MSE	
ABC	2	SSABC	SSABC/2	MSABC/MSE	
Error	55	SSE	SSE/55		
Total	71	SSTot			

15.33
  a. Randomized Complete Block Design with the 5 specimens of fabrics serving as the blocks and the three dyes being the treatments.
  b. The tests for the differences in mean quality of the three dyes has p-value=0.01. Thus, there is significant evidence of a difference in the mean quality of the three dyes.
  Using Tukey's W-procedure with $\alpha = 0.05, s_\epsilon^2 = MSE = 34.367, q_\alpha(t, df_{Error}) = q_\alpha(3,8) = 4.04 \Rightarrow$

$$W = (4.04)\sqrt{\frac{34.367}{5}} = 10.59 \Rightarrow$$

	Dye		
	A	B	C
Mean	77.40	84.60	92.80
Grouping	a	ab	b

  c. $t = 3, b = 5 \Rightarrow RE(RCB, CR) = \frac{(5-1)(23.567)+(5)(3-1)(34.367)}{((5)(3)-1)(34.367)} = 0.91 \Rightarrow$
  It would take 0.91 times as many observations (approximately 5) per treatment in a completely randomized design to achieve the same level of precision in estimating the treatment means as was accomplished in the randomized complete block design. Since RE was slightly less than 1, we would conclude that the blocking was not effective.

15.35
  a. Latin Square Design with blocking variables Farm and Fertility. The treatment is the five types of fertilizers.
  b. There is significant evidence (p-value<0.0001) the mean yields are different for the five fertilizers.

15.37
  a. This is a generalized randomized block design with center as the blocking factor and 2 factors of interest (exercise regimen and age group).
  b. A model for this experiment is given here:
  $y_{ijk} = \mu + v_i + \alpha_j + \beta_k + \alpha\beta_{jk} + \epsilon_{ijk}$;
  with $i = 1, 2, 3, 4, 5; j = 1, 2, 3; k = 1, 2, 3$;
  where y_{ijk} is the reduction in blood cholesterol of the ith center with the jth exercise regimen, and k^{th} age group.
  v_i is the effect of the ith center on reduction in blood cholesterol
  α_j is the effect of the jth exercise regimen on reduction in blood cholesterol
  β_k is the effect of the kth age group on reduction in blood cholesterol
  $\alpha\beta_{jk}$ is the interaction effect of the jth exercise regimen and k^{th} age group on reduction in blood cholesterol

c. The AOV table is given below.

```
Analysis of Variance

Source          DF    Adj SS    Adj MS   F-Value   P-Value
  Center         4     30690   7672.57     13.83     0.000
  Exercise       2      3763   1881.43      3.39     0.037
  Age            2      2773   1386.36      2.50     0.086
  Exercise*Age   4       168     42.03      0.08     0.989
Error          122     67671    554.68
  Lack-of-Fit   32     11297    353.04      0.56     0.966
  Pure Error    90     56373    626.37
Total          134    105065
```

Tukey Pairwise Comparisons: Response = Reduction, Term = Exercise

```
Grouping Information Using the Tukey Method and 95% Confidence

Exercise   N     Mean    Grouping
90        45   16.4222   A
60        45    6.4000   A    B
30        45    4.3333        B

Means that do not share a letter are significantly different.
```

There is a significant difference in mean reduction across exercise regimens. The Tukey tests show that mean reduction in blood cholesterol for the 90 min regimen is significantly lower than that of the 30 min regimen.

d. The interaction term is non-significant so the results in part c can be applied to each of the age groups.

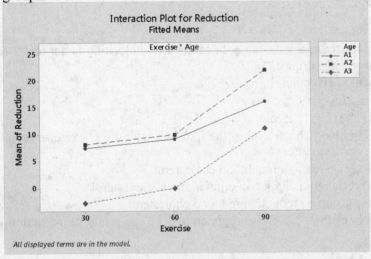

e. Because there is no significant interaction, the groupings developed in part c are applicable here.

15.39
a. The residual plot is shown below. There are no violations of the assumptions.

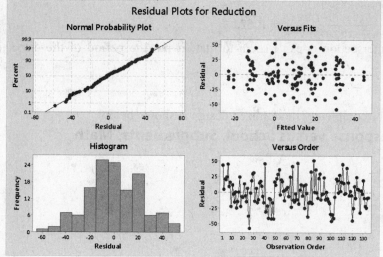

b. We can use the Friedman's test. But, because this test can only be used for RCBD and not when there is more than observation per treatment within a block, the analysis was run for the mean response per treatment within each block (that is, responses were averaged over that three observations per treatment, per Center.)

As can be seen from the output from Minitab, there is no evidence (p-value = 0.144) of a significant difference between the age and exercise regimen combinations.

```
Friedman Test: Response versus Trt blocked by Center

S = 12.12   DF = 8   P = 0.146
S = 12.16   DF = 8   P = 0.144 (adjusted for ties)

         Est    Sum of
Trt  N  Median  Ranks
 1   5  19.67   33.0
 2   5  31.02   35.0
 3   5  14.78   29.0
 4   5  14.39   24.5
 5   5  15.32   29.0
 6   5   5.74   14.0
 7   5  12.30   23.0
 8   5  14.24   24.0
 9   5   3.22   13.5

Grand median = 14.52
```

c. Because there is no reason to suspect violation of any assumptions, we can use the normal assumption based conclusions from 15.37.

15.41

a. $RE(RCB, CR) = \frac{(5-1)4.557+5(5-1)5.841}{[(5)(5)-1](5.841)} = 0.9633$

b. $RE(RCB, CR) = \frac{(5-1)272.917+5(5-1)5.841}{[(5)(5)-1](5.841)} = 8.621$

c. In future studies, the intersection doesn't seem to matter, but the period of the day should be controlled for.

15.43

a. The AOV results with the completely randomized design is shown below.

General Linear Model: Response versus School, Supplements, Math

```
Method

Factor coding   (-1, 0, +1)


Factor Information

Factor        Type    Levels  Values
School        Fixed        2  HighSch, JunHigh
Supplements   Fixed        2  No, Yes
Math          Fixed        3  Hgh, Low, Med

Analysis of Variance

Source                      DF   Adj SS    Adj MS  F-Value  P-Value
  School                     1   588.06   588.060    17.11    0.001
  Supplements                1   719.41   719.415    20.93    0.001
  Math                       2  1042.36   521.179    15.17    0.001
  School*Supplements         1    54.00    54.000     1.57    0.234
  School*Math                2    30.54    15.271     0.44    0.651
  Supplements*Math           2    33.94    16.969     0.49    0.622
  School*Supplements*Math    2    11.07     5.536     0.16    0.853
Error                       12   412.40    34.367
Total                       23  2891.78


Model Summary

      S    R-sq  R-sq(adj)  R-sq(pred)
5.86231  85.74%     72.67%      42.96%


Coefficients

Term                  Coef  SE Coef  T-Value  P-Value   VIF
Constant             29.27     1.20    24.46    0.000
School
  HighSch             4.95     1.20     4.14    0.001  1.00
Supplements
  No                 -5.47     1.20    -4.58    0.001  1.00
Math
  Hgh                 6.79     1.69     4.01    0.002  1.33
  Low                -8.93     1.69    -5.27    0.000  1.33
School*Supplements
  HighSch No         -1.50     1.20    -1.25    0.234  1.00
```

```
School*Math
  HighSch Hgh                        0.04      1.69     0.02    0.983   1.33
  HighSch Low                       -1.40      1.69    -0.83    0.424   1.33
Supplements*Math
  No Hgh                            -1.19      1.69    -0.70    0.496   1.33
  No Low                             1.63      1.69     0.96    0.356   1.33
School*Supplements*Math
  HighSch No Hgh                    -0.29      1.69    -0.17    0.868   1.33
  HighSch No Low                    -0.65      1.69    -0.38    0.708   1.33


Regression Equation

Response = 29.27 + 4.95 School_HighSch - 4.95 School_JunHigh - 5.47 Supplements_No
           + 5.47 Supplements_Yes + 6.79 Math_Hgh - 8.93 Math_Low + 2.14 Math_Med
           - 1.50 School*Supplements_HighSch No + 1.50 School*Supplements_HighSch Yes
           + 1.50 School*Supplements_JunHigh No - 1.50 School*Supplements_JunHigh Yes
           + 0.04 School*Math_HighSch Hgh - 1.40 School*Math_HighSch Low
           + 1.36 School*Math_HighSch Med - 0.04 School*Math_JunHigh Hgh
           + 1.40 School*Math_JunHigh Low - 1.36 School*Math_JunHigh Med
           - 1.19 Supplements*Math_No Hgh + 1.63 Supplements*Math_No Low
           - 0.44 Supplements*Math_No Med + 1.19 Supplements*Math_Yes Hgh
           - 1.63 Supplements*Math_Yes Low + 0.44 Supplements*Math_Yes Med
           - 0.29 School*Supplements*Math_HighSch No Hgh
           - 0.65 School*Supplements*Math_HighSch No Low
           + 0.94 School*Supplements*Math_HighSch No Med
           + 0.29 School*Supplements*Math_HighSch Yes Hgh
           + 0.65 School*Supplements*Math_HighSch Yes Low
           - 0.94 School*Supplements*Math_HighSch Yes Med
           + 0.29 School*Supplements*Math_JunHigh No Hgh
           + 0.65 School*Supplements*Math_JunHigh No Low
           - 0.94 School*Supplements*Math_JunHigh No Med
           - 0.29 School*Supplements*Math_JunHigh Yes Hgh
           - 0.65 School*Supplements*Math_JunHigh Yes Low
           + 0.94 School*Supplements*Math_JunHigh Yes Med
```

    The tests can all be run. All main effects are significant, but no interactions are significant.

b.   The only question that arises is how students of high, med, and low math ability were assigned to specific classrooms.

Chapter 16

The Analysis of Covariance

16.1
$$y_i = \beta_0 + \beta_1 x_{1i} + \beta_2 x_{2i} + \beta_3 x_{3i} + \beta_4 x_{4i} + \\ \beta_5 x_{1i} x_{2i} + \beta_6 x_{1i} x_{3i} + \beta_7 x_{1i} x_{4i} + \epsilon_i \text{ for } i = 1, \ldots, 80$$

x_1 = age

$x_2 = \begin{cases} 1 & \text{if Supplement 1 is used} \\ 0 & \text{Otherwise} \end{cases}$

$x_3 = \begin{cases} 1 & \text{if Supplement 2 is used} \\ 0 & \text{Otherwise} \end{cases}$

$x_4 = \begin{cases} 1 & \text{if Supplement 3 is used} \\ 0 & \text{Otherwise} \end{cases}$

The coefficients are identified through the mean response under each treatment

Treatment	Mean Response
Control	$\beta_0 + \beta_1 x_1$
Supplement 1	$(\beta_0 + \beta_2) + (\beta_1 + \beta_5) x_1$
Supplement 2	$(\beta_0 + \beta_3) + (\beta_1 + \beta_6) x_1$
Supplement 3	$(\beta_0 + \beta_4) + (\beta_1 + \beta_7) x_1$

16.3

a. In order to test for differences in the adjusted treatment means, two models needed to be fitted to test:
H_0: No Difference in Adjusted Treatment Means
H_a: Difference in Adjusted Treatment Means
In terms of model parameters, we have:
$H_0: \beta_2 = \beta_3 = \beta_4 = 0$.
H_a: At least one of $\beta_2, \beta_3, \beta_4$ is not zero

Model 1 (Assumes H_a is true):
$y_i = \beta_0 + \beta_1 x_{1i} + \beta_2 x_{2i} + \beta_3 x_{3i} + \beta_4 x_{4i} + \epsilon_i$

Model 2 (Assumes H_0 is true): $y_i = \beta_0 + \beta_1 x_{1i} + \epsilon_i$

$$MS_{Drop} = \frac{SSE_2 - SSE_1}{df_{E2} - df_{E1}} = \frac{SSE_2 - SSE_1}{(80-2)-(80-5)}$$

$F = \frac{MS_{Drop}}{MSE_1}$ with $df_1 = ((80-2)-(80-5)) = 3$ and $df_2 = 80 - 5 = 75$

b. For supplement 1 with $x_1 = 45$: $\hat{y} = \hat{\beta}_0 + \hat{\beta}_1(45) + \hat{\beta}_2$

16.5
a. $y_i = \beta_0 + \beta_1 x_{1i} + \beta_2 x_{2i} + \beta_3 x_{3i} + \beta_4 x_{1i} x_{2i} + \beta_5 x_{1i} x_{3i} + \epsilon_i$ i=1,...30

x_1 = Number of Cigarettes

$x_2 = \begin{cases} 1 & \text{if Treatment I is applied} \\ 0 & \text{if Otherwise} \end{cases}$

$x_3 = \begin{cases} 1 & \text{if Treatment II is applied} \\ 0 & \text{if Otherwise} \end{cases}$

The coefficients are identified through the mean response under each treatment

Treatment	Mean Response
C	$\beta_0 + \beta_1 x_1$
I	$(\beta_0 + \beta_2) + (\beta_1 + \beta_4) x_1$
II	$(\beta_0 + \beta_3) + (\beta_1 + \beta_5) x_1$

b. The scatterplot is shown below:

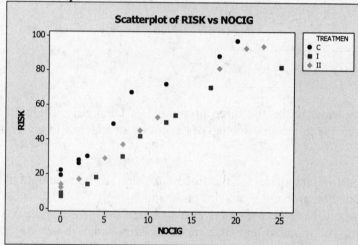

It appears there is a relationship between risk and number of cigarettes smoked.

c. The slopes in part b look the same. Parallelism appears to be a reasonable assumption to make.

16.7
a. $y_i = \beta_0 + \beta_1 x_{1i} + \beta_2 x_{2i} + \beta_3 x_{3i} + \beta_4 x_{1i} x_{2i} + \beta_5 x_{1i} x_{3i} + \epsilon_i$ i=1,...30

x_1 = average monthly sales for the 12 months prior to promotion

$x_2 = \begin{cases} 1 & \text{if Promotion B is used} \\ 0 & \text{if Otherwise} \end{cases}$

$x_3 = \begin{cases} 1 & \text{if Promotion C is used} \\ 0 & \text{if Otherwise} \end{cases}$

214 Chapter 16: The Analysis of Covariance

The coefficients are identified through the mean response under each treatment

Promotion	Mean Response
A	$\beta_0 + \beta_1 x_1$
B	$(\beta_0 + \beta_2) + (\beta_1 + \beta_4)x_1$
C	$(\beta_0 + \beta_3) + (\beta_1 + \beta_5)x_1$

b. The scatterplot is shown below:

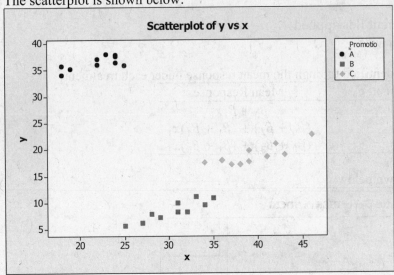

c. The slopes appear to be the same under the three promotions. Obviously, the extrapolation problem appears to be an issue due to the non-overlap of the covariate under differing groups.

16.9
a. Yes. The covariate covers a different space for the three variables so the assumptions are not wholly satisfied. The extrapolation problem is a major problem.
b. To avoid this problem, it would have been a good idea to block on the covariate x (average monthly sales in the previous 12 months) and fit a randomized complete block design. This would cause the treatments (promotions) to be randomly applied to the whole range of covariates.

16.11
a.
$$y_i = \beta_0 + \beta_1 x_{1i} + \beta_2 x_{2i} + \beta_3 x_{3i} + \beta_4 x_{4i} + \beta_5 x_{5i} + \beta_6 x_{1i}x_{2i} + \beta_7 x_{1i}x_{3i} + \beta_8 x_{1i}x_{4i} + \beta_9 x_{1i}x_{5i}$$
$$+ \beta_{10} x_{6i} + \beta_{11} x_{7i} + \beta_{12} x_{8i} + \beta_{13} x_{1i}^2 + \beta_{14} x_{1i}^2 x_{2i} + \beta_{15} x_{1i}^2 x_{3i} + \beta_{16} x_{1i}^2 x_{4i}$$
$$+ \beta_{17} x_{1i}^2 x_{5i} + \epsilon_i \text{ for } i = 1, \ldots, 20$$

x_1 = covariate

$$x_2 = \begin{cases} 1 & \text{if Treatment 2 is applied} \\ 0 & \text{if Otherwise} \end{cases}$$

$$x_3 = \begin{cases} 1 & \text{if Treatment 3 is applied} \\ 0 & \text{if Otherwise} \end{cases}$$

$$x_4 = \begin{cases} 1 & \text{if Treatment 4 is applied} \\ 0 & \text{if Otherwise} \end{cases}$$

$$x_5 = \begin{cases} 1 & \text{if Treatment 5 is applied} \\ 0 & \text{if Otherwise} \end{cases}$$

$$x_6 = \begin{cases} 1 & \text{if Observation in Block 2} \\ 0 & \text{if Otherwise} \end{cases}$$

$$x_7 = \begin{cases} 1 & \text{if Observation in Block 3} \\ 0 & \text{if Otherwise} \end{cases}$$

$$x_8 = \begin{cases} 1 & \text{if Observation in Block 4} \\ 0 & \text{if Otherwise} \end{cases}$$

b. Parallelism results from the five lines having the same slope for linear and quadratic terms, i.e., the correct model follows the null hypothesis:
$H_0: \beta_6 = \beta_7 = \beta_8 = \beta_9 = \beta_{14} = \beta_{15} = \beta_{16} = \beta_{17} = 0$.
A test of the alternative
H_a: At least one of $\beta_6, \beta_7, \beta_8, \beta_9, \beta_{14}, \beta_{15}, \beta_{16}, \beta_{17}$ is not zero
Consists of fitting two models:
Model 1 (Assumes H_a is true): Full Model
Model 2 (Assumes H_0 is true): $y_i = \beta_0 + \beta_1 x_{1i} + \beta_2 x_{2i} + \beta_3 x_{3i} + \beta_4 x_{4i} + \beta_5 x_{5i} + \beta_{10} x_{6i} + \beta_{11} x_{7i} + \beta_{12} x_{8i} + \beta_{13} x_{1i}^2 + \epsilon_i$

$$MS_{Drop} = \frac{SSE_2 - SSE_1}{df_{E2} - df_{E1}} = \frac{SSE_2 - SSE_1}{(20-10)-(20-18)}$$
$F = \frac{MS_{Drop}}{MSE_1}$ with $df_1 = ((20-10)-(20-18)) = 8$ and $df_2 = 20 - 18 = 2$

c. $df_1 = 8$ and $df_2 = 2$

d. In terms of model parameters, we have:
$H_0: \beta_2 = \beta_3 = \beta_4 = \beta_5 = 0$.
H_a: At least one of $\beta_2, \beta_3, \beta_4, \beta_5$ is not zero

Model 1: (Assumes H_a is true): $y_i = \beta_0 + \beta_1 x_{1i} + \beta_2 x_{2i} + \beta_3 x_{3i} + \beta_4 x_{4i} + \beta_5 x_{5i} + \beta_{10} x_{6i} + \beta_{11} x_{7i} + \beta_{12} x_{8i} + \beta_{13} x_{1i}^2 + \epsilon_i$

Model 2 (Assumes H_0 is true): $y_i = \beta_0 + \beta_1 x_{1i} + \beta_{10} x_{6i} + \beta_{11} x_{7i} + \beta_{12} x_{8i} + \beta_{13} x_{1i}^2 + \epsilon_i$

$$MS_{Drop} = \frac{SSE_2 - SSE_1}{df_{E2} - df_{E1}} = \frac{SSE_2 - SSE_1}{(20-6)-(20-10)} = \frac{SSE_2 - SSE_1}{4}$$
$F = \frac{MS_{Drop}}{MSE_1}$ with $df_1 = ((20-6)-(20-10)) = 4$ and $df_2 = 20 - 10 = 10$

16.13

a. The ANCOVA table is shown below.

```
Analysis of Variance

Source   DF   Adj SS    Adj MS    F-Value   P-Value
  GL      1   0.06498   0.064979   15.94    0.001
  Batch   5   0.09761   0.019522    4.79    0.005
  IMT     4   0.25403   0.063507   15.58    0.000
Error    19   0.07744   0.004076
Total    29   4.43059
```

There is a significant difference in the immersion levels mean DVBN after accounting for glycogen levels. There is also a significant difference in DVBN among the batches.

b. The prediction is shown here. (Since we didn't know the batch—block—that term was omitted for the prediction.)

Prediction for DVBN

```
General Linear Model Information

Terms
GL IMT


Variable   Setting
GL            3
IMT          20


  Fit      SE Fit        95% CI              95% PI
17.1323   0.0354376  (17.0591, 17.2054)  (16.9414, 17.3231)
```

16.15

a. Randomized complete block design with the three antidepressants as treatments, age-sex combinations as six blocks, and the pretreatment rating as a covariate.

b. $y_i = \beta_0 + \beta_1 x_{1i} + \beta_2 x_{2i} + \beta_3 x_{3i} + \beta_4 x_{1i} x_{2i} + \beta_5 x_{1i} x_{3i} + \beta_6 x_{4i} + \beta_7 x_{5i} + \beta_8 x_{6i} + \beta_9 x_{7i} + \beta_{10} x_{8i} + \epsilon_i$ for $i = 1, \ldots, 16$

c.
$x_1 = $ pretreatment

$x_2 = \begin{cases} 1 & \text{if} \quad \text{Antidepressant B} \\ 0 & \text{if} \quad \text{Otherwise} \end{cases}$

$x_3 = \begin{cases} 1 & \text{if} \quad \text{Antidepressant C} \\ 0 & \text{if} \quad \text{Otherwise} \end{cases}$

$x_4 = \begin{cases} 1 & \text{if} \quad \text{Observation in Block 2} \\ 0 & \text{if} \quad \text{Otherwise} \end{cases}$

$x_5 = \begin{cases} 1 & \text{if} \quad \text{Observation in Block 3} \\ 0 & \text{if} \quad \text{Otherwise} \end{cases}$

$$x_6 = \begin{cases} 1 & \text{if} \quad \text{Observation in Block 4} \\ 0 & \text{if} \quad \text{Otherwise} \end{cases}$$

$$x_7 = \begin{cases} 1 & \text{if} \quad \text{Observation in Block 5} \\ 0 & \text{if} \quad \text{Otherwise} \end{cases}$$

$$x_8 = \begin{cases} 1 & \text{if} \quad \text{Observation in Block 6} \\ 0 & \text{if} \quad \text{Otherwise} \end{cases}$$

16.17
a. Test for difference in adjusted means between blocks:
$$F = \frac{(75.217 - 68.581)/(14 - 9)}{68.581/9} = 0.17, \text{ with df=5,9} \Rightarrow$$
$p - value = \Pr(F_{5,9} \geq 0.17) = 0.966$
There is not significant evidence that the adjusted mean ratings are different for the six blocks.
b. The sum of squares associated the block indicator variables: x_4, x_5, x_6, x_7, x_8 partition the sum of squares for blocks into five separate sums of squares, a sum of squares for each indicator variable.
c. Because there was not a significant block effect, none of the five individual sum of squares would be significant.

16.19
a. Test for parallelism of the three treatment lines:
$$F = \frac{(3316.8281 - 3180.7299)/(75 - 72)}{3180.7299/72} = 1.03, \text{ with df=3,72} \Rightarrow$$
$p - value = \Pr(F_{3,72} \geq 1.03) = 0.3859 \Rightarrow$
There is not significant evidence that the lines are not parallel.
b. Test for difference in adjusted treatment means:
$$F = \frac{(8724.7852 - 3316.8281)/(78 - 75)}{3316.8281/75} = 40.76, \text{ with df=3,75} \Rightarrow$$
$p - value = \Pr(F_{3,75} \geq 40.76) < 0.0001 \Rightarrow$
There is significant evidence that the adjusted mean ratings are different for the four socioeconomic classes.
c. $\hat{\mu}_{adj,1} = (\hat{\beta}_0 + \hat{\beta}_2) + \hat{\beta}_1 \bar{x}_{..} = (37.197 - 22.490) + (0.27472)(28.95) = 22.66$
$\hat{\mu}_{adj,2} = (\hat{\beta}_0 + \hat{\beta}_3) + \hat{\beta}_1 \bar{x}_{..} = (37.197 - 15.951) + (0.27472)(28.95) = 29.20$
$\hat{\mu}_{adj,3} = (\hat{\beta}_0 + \hat{\beta}_4) + \hat{\beta}_1 \bar{x}_{..} = (37.197 - 14.784) + (0.27472)(28.95) = 30.37$
$\hat{\mu}_{adj,4} = \hat{\beta}_0 + \hat{\beta}_1 \bar{x}_{..} = 37.197 + (0.27472)(28.95) = 45.15$

$SE(\hat{\mu}_{adj,1}) = \sqrt{MSE\left(\frac{1}{n_i} + \frac{(\bar{x}_{1.} - \bar{x}_{..})^2}{E_{xx}}\right)} = \sqrt{(44.2244)\left(\frac{1}{20} + \frac{(28.95-28.95)^2}{9135.8}\right)} = 1.4870$

$SE(\hat{\mu}_{adj,2}) = \sqrt{MSE\left(\frac{1}{n_i} + \frac{(\bar{x}_{1.} - \bar{x}_{..})^2}{E_{xx}}\right)} = \sqrt{(44.2244)\left(\frac{1}{20} + \frac{(28.70-28.95)^2}{9135.8}\right)} = 1.4871$

$SE(\hat{\mu}_{adj,3}) = \sqrt{MSE\left(\frac{1}{n_i} + \frac{(\bar{x}_{1.} - \bar{x}_{..})^2}{E_{xx}}\right)} = \sqrt{(44.2244)\left(\frac{1}{20} + \frac{(28.60-28.95)^2}{9135.8}\right)} = 1.4872$

$SE(\hat{\mu}_{adj,4}) = \sqrt{MSE\left(\frac{1}{n_i} + \frac{(\bar{x}_{1.} - \bar{x}_{..})^2}{E_{xx}}\right)} = \sqrt{(44.2244)\left(\frac{1}{20} + \frac{(29.55-28.95)^2}{9135.8}\right)} = 1.4876$

The Bonferroni $\alpha = \frac{0.05}{(2)(4)} = 0.00625 \Rightarrow t_{0.00625,75} = 2.559$

95% C.I.'s for the mean adjusted verbalization scores:
Socioeconomic Class 1: $22.66 \pm (2.559)(1.4870) \Rightarrow (18.8, 26.5)$
Socioeconomic Class 2: $29.20 \pm (2.559)(1.4871) \Rightarrow (25.4, 33.0)$
Socioeconomic Class 3: $30.37 \pm (2.559)(1.4872) \Rightarrow (26.6, 34.2)$
Socioeconomic Class 4: $45.15 \pm (2.559)(1.4876) \Rightarrow (41.3, 49.0)$

16.21
a. Test for parallelism of the three treatment lines:
$$F = \frac{(1213.1 - 978.6)/(26 - 24)}{978.6/24} = 2.88, \text{ with df}=2,24 \Rightarrow$$
$p - value = \Pr(F_{2,24} \geq 2.88) = 0.076 > 0.05 \Rightarrow$
There is not significant evidence that the lines are not parallel.

b. Test for difference in adjusted treatment means:
$$F = \frac{(9899.9 - 1213.1)/(28 - 26)}{1213.1/26} = 93.1, \text{ with df}=2,26 \Rightarrow$$
$p - value = \Pr(F_{2,26} \geq 99.67) < 0.0001 \Rightarrow$
There is significant evidence that the adjusted mean ratings are different for the three processes.

c.
```
Term          Coef    SE Coef      T       P
Constant     -7.62     11.44    -0.67   0.511
x           4.2026    0.3725    11.28   0.000
PROCESS
P1           -2.660    1.764    -1.51   0.144
P2           22.081    1.766    12.50   0.000
```

$\hat{\mu}_{adj,1} = (\hat{\beta}_0 + \hat{\beta}_3) + \hat{\beta}_1 \bar{x}_{..} = (-7.62 - 2.66) + (4.2026)(30.53) = 118.03$
$\hat{\mu}_{adj,2} = (\hat{\beta}_0 + \hat{\beta}_4) + \hat{\beta}_1 \bar{x}_{..} = (-7.62 + 22.081) + (4.2026)(30.53) = 142.77$
$\hat{\mu}_{adj,3} = \hat{\beta}_0 + \hat{\beta}_1 \bar{x}_{..} = -27.041 + (4.2026)(30.53) = 101.26$

$SE(\hat{\mu}_{adj,1}) = \sqrt{MSE\left(\frac{1}{n_i} + \frac{(\bar{x}_{1.} - \bar{x}_{..})^2}{E_{xx}}\right)} = \sqrt{(46.7)\left(\frac{1}{30} + \frac{(30.5-30.53)^2}{337.413}\right)} = 1.248$

$SE(\hat{\mu}_{adj,2}) = \sqrt{MSE\left(\frac{1}{n_i} + \frac{(\bar{x}_{1.} - \bar{x}_{..})^2}{E_{xx}}\right)} = \sqrt{(46.7)\left(\frac{1}{30} + \frac{(30.3-30.53)^2}{337.413}\right)} = 1.251$

$SE(\hat{\mu}_{adj,3}) = \sqrt{MSE\left(\frac{1}{n_i} + \frac{(\bar{x}_{1.} - \bar{x}_{..})^2}{E_{xx}}\right)} = \sqrt{(46.7)\left(\frac{1}{30} + \frac{(30.8-30.53)^2}{337.413}\right)} = 1.252$

The Bonferroni $\alpha = \frac{0.05}{(2)(3)} = 0.00833 \Rightarrow t_{0.00833,26} = 2.559$

95% C.I.'s for the mean adjusted verbalization scores:
Process 1: $118.03 \pm (2.559)(1.248) \Rightarrow (114.84, 121.22)$
Process 2: $142.77 \pm (2.559)(1.251) \Rightarrow (139.57, 145.97)$
Process 3: $101.26 \pm (2.559)(1.252) \Rightarrow (98.06, 104.46)$

Process 2 differs significantly from the others.

16.23
  a. The ANCOVA table is shown below.
```
Analysis of Variance

Source         DF   Adj SS   Adj MS  F-Value  P-Value
  Ppt           1    36.69    36.69     0.80    0.378
  Region        4   203.45    50.86     1.10    0.369
  Age           2   231.99   116.00     2.52    0.094
Error          37  1704.89    46.08
  Lack-of-Fit  35  1656.43    47.33     1.95    0.396
  Pure Error    2    48.45    24.23
Total          44  2760.36
```
There is not a significant difference in mean Fisher's alpha across the age groups after accounting for annual precipitation.

  b. The Tukey results are shown below.

Tukey Pairwise Comparisons: Response = FisherAlpha, Term = Age

```
Grouping Information Using the Tukey Method and 95% Confidence

Age   N    Mean  Grouping
 3   14  29.5255  A
 1   16  28.6904  A
 2   15  23.5956  A

Means that do not share a letter are significantly different.
```

  c. The residual plot is shown below.

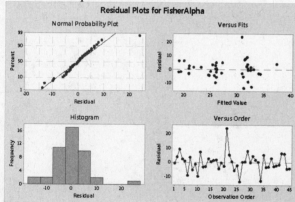

The only potential violation is the high residual outlier in observation 21.

16.25 Results are consistent when considering the age and/or topography groups (no significant difference in mean Fisher alpha or Shannon Index among age and or topography groups). The benefit to Shannon index is the lack of a large (outlier) value in observation 21.

16.27
  a. $F - alpha = \mu + A_i + T_j + AT_{ij} + \beta_1 P + \varepsilon_{ijk}$
  b. The analysis using the data in this form is impossible because there are no topography 1 or 2 for age 3
  c. To complete the analysis, I grouped age groups 1 and 2 into a single group.

```
Analysis of Variance

Source           DF   Adj SS   Adj MS   F-Value  P-Value
  Ppt             1   109.61   109.61    2.19    0.148
  Region          4   262.45    65.61    1.31    0.286
  Topography      2    64.55    32.28    0.64    0.531
  Age2            1   142.83   142.83    2.85    0.101
  Topography*Age2 2    88.33    44.17    0.88    0.423
Error            34  1703.78    50.11
  Lack-of-Fit    32  1655.33    51.73    2.14    0.370
  Pure Error      2    48.45    24.23
Total            44  2760.36
```

Results are still consistent. No difference among age groups, topography groups, or the interaction.

Chapter 17

Analysis of Variance for Some Fixed-, Random-, and Mixed-Effects Models

17.1
a. $y_{ij} = \mu + \alpha_i + \epsilon_{ij}$; i=1,...10; j=1,2,3,4,5 where y_{ij} is the percentage of ingredient in the paint for the jth can in the ith batch

μ is the mean percentage of the ingredient in the paint
α_i is a random effect due to the ith batch
ε_{ij} is the random effect due to all other sources but batch

b. The Minitab output is given below:
General Linear Model: Percenta versus Batch

```
Factor   Type     Levels  Values
Batch    random       10  1, 2, 3, 4, 5, 6, 7, 8, 9, 10

Analysis of Variance for Percenta, using Adjusted SS for Tests

Source  DF   Seq SS   Adj SS   Adj MS      F      P
Batch    9   51.643   51.643    5.738   1.26  0.289
Error   40  182.338  182.338    4.558
Total   49  233.981

S = 2.13505   R-Sq = 22.07%   R-Sq(adj) = 4.54%
```

From the output, the p-value for the significance of batch is p-value = 0.289 > 0.05. This means there is not significant evidence that the batches exhibit different average percentages of the ingredient.

c. $\hat{\sigma}_\varepsilon^2 = MSE$ and $\hat{\sigma}_\tau^2 = (MST - MSE)/n$ so
$\hat{\sigma}_\varepsilon^2 = 4.558$
$\hat{\sigma}_\tau^2 = \dfrac{(5.738 - 4.558)}{5} = 0.236$

Proportion $= \dfrac{\hat{\sigma}_\tau^2}{\hat{\sigma}_\tau^2 + \hat{\sigma}_\varepsilon^2} = \dfrac{0.236}{0.236 + 4.558} = 4.9\%$

17.3
a. The following model is applicable to both scenarios. The difference is in the interpretation of parameters.
$y_{ij} = \mu + \alpha_i + \epsilon_{ij}$; i=1,2,3,4,5; j=1,2,3,4,5 where y_{ij} is the average daily weight gain of calves sired by Bull i

Scenario A:
μ is the mean daily weight gain of all calves
α_i is the fixed effect due to the ith bull (sire)
ϵ_{ij} is the random effect due to all other sources but bull

Scenario B:
μ is the mean daily weight gain of all calves
α_i is the random effect due to the ith bull (sire)
ϵ_{ij} is the random effect due to all other sources but bull

b. Scenario A:
$$H_0: \alpha_1 = \alpha_2 = \alpha_3 = \alpha_4 = \alpha_5 = 0 \text{ versus } H_a: \text{at least one } \alpha_i \neq 0$$
Scenario B:
$$H_0: \sigma_\alpha^2 = 0 \text{ versus } H_a: \sigma_\alpha^2 > 0 \text{ where } \sigma_\alpha^2 \text{ is the variability of the sires}$$

17.5
a. Only females of essentially the same age were selected to remove variability due to gender or age that may enter the model. Rather than add additional parameters, the experimenter chose to simply keep the gender/age constant so no blocks need to be added. The additional variability in gender/age could artificially inflate the random variability among subjects and dilute the treatment effect.

b. $y_{ij} = \mu + \tau_i + \beta_j + \epsilon_{ij}$; i=1,...,10; j=1,2,3,4,5
where y_{ij} is the average DNA content in the plaque of subject i analyzed by analyst j
μ is the mean plaque DNA content of all subjects/analysts
τ_i is the random effect of the ith subject
β_j is the random effect of the jth analyst
ϵ_{ij} is the random effect due to all other sources

c. $H_0: \sigma_A^2 = 0$ versus $H_a: \sigma_A^2 > 0$ where σ_A^2 is the variability of the analysts

17.7
a. $y_{ijk} = \mu + \tau_i + \beta_j + \tau\beta_{ij} + \epsilon_{ijk}$; i=1,...,10; j=1,...,6; k=1,2
where y_{ijk} is the number of microorganisms in the kth beer under the ith process analyzed by the jth lab
μ is the mean number of microorganisms over all labs/processes
τ_i is the random effect of the ith process
β_j is the random effect of the jth lab
$\tau\beta_{ij}$ is the random interaction effect of the ith process and j^{th} lab
ϵ_{ijk} is the random effect due to all other sources

b. The table of expected mean squares is shown below

Source	Expected Mean Squares
Process	$\sigma_\epsilon^2 + 2\sigma_{\tau\beta}^2 + 12\sigma_\tau^2$
Lab	$\sigma_\epsilon^2 + 2\sigma_{\tau\beta}^2 + 20\sigma_A^2$
Process*Lab	$\sigma_\epsilon^2 + 2\sigma_{\tau\beta}^2$
Error	σ_ϵ^2

c. To test for an interaction effect:
$H_0: \sigma_{\tau\beta}^2 = 0$ versus $H_a: \sigma_{\tau\beta}^2 > 0$
To test for a process effect:
$H_0: \sigma_\tau^2 = 0$ versus $H_a: \sigma_\tau^2 > 0$
To test for a lab effect:
$H_0: \sigma_\beta^2 = 0$ versus $H_a: \sigma_\beta^2 > 0$

17.9
a. If the factors are fixed, their levels are chosen before the experiment as the only levels of interest. If they are random factors, the treatments (or levels) are randomly selected from a population of possible treatments (levels).
b. When the factors are fixed, inference can only be used on those levels in the design. When the factors are random, the treatment levels are chosen from a population of possible treatments and inferences can therefore be extended to all treatments in the population, not simply those in the experimental design.
c. Factor A inferences are only relevant to the levels chosen in the experiment. Factor B inferences apply to all treatments in the population.

17.11
a. The AOV table is given here:
```
Source              DF   Adj SS   Adj MS   F-Value   P-Value
  Location           4    3.811   0.9529    0.71     0.602
  Chemical           3  180.133  60.0443   44.59     0.000
  Location*Chemical 12   16.159   1.3465    3.89     0.004
Error                20    6.925  0.3462
Total                39  207.028
```

b. The F-test for $H_0: \sigma_{\alpha\beta}^2 = 0$ versus $H_a: \sigma_{\alpha\beta}^2 > 0$ has p-value = 0.004. Therefore, there is significant evidence of an interaction between Locations and Chemicals. This means that the efficacy of a particular chemical depends on the location.

c. The F-test for $H_0: \sigma_\alpha^2 = 0$ versus $H_a: \sigma_\alpha^2 > 0$ has p-value = 0.601. Therefore, there is not significant evidence of an effect due to Locations.

The F-test for $H_0: \beta_1=...=\beta_4=0$ versus H_a: at least one $\beta_i \neq 0$ has p-value < 0.0001. Therefore, there is significant evidence of an effect due to Chemicals.

d. The chemicals efficacy differs based on the location.
e. The proportion of variation in the number of fire ants killed attributed to the sources in the model is $R^2 = 86.62\%$. The breakdown of the variance component is shown below.

224 Chapter 17: Analysis of Variance for Some Fixed-, Random-, and Mixed-Effects Models

```
Variance Components, using Adjusted SS

Source                Variance    % of Total     StDev    % of Total
Location            -0.0492083*        0.00%   0.000000        0.00%
Location*Chemical     0.500146        59.09%   0.707210       76.87%
Error                 0.34625         40.91%   0.588430       63.96%
Total                 0.846396                 0.919998
```

f. The residual plot shows no major violations of assumptions.

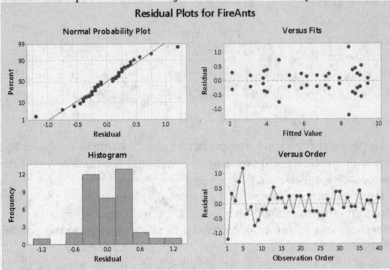

17.13
a. The model for a 5x6x3 factorial treatment structure with $n = 10$ replications and factors A aqnd B random and factor fixed is as follows:

$$y_{ijkm} = \mu + \alpha_i + \beta_j + \gamma_k + \alpha\beta_{ij} + \alpha\gamma_{ik} + \beta\gamma_{jk} + \alpha\beta\gamma_{ijk} + \epsilon_{ijkm}; i = 1,\ldots,5; j = 1,\ldots,6; k = 1,2,3; m = 1,\ldots,10$$

Where y_{ijkm} is the response of the mth replication under level i factor A, level j factor B, level k factor C.

μ is the mean of all responses
α_i is the random effect of the ith level of factor A
β_j is the random effect of the jth level of factor B
γ_k is the fixed effect of the kth level of factor C
$\alpha\beta_{ij}$ is the random interaction effect of the ith level of factor A with the jth level of factor B
$\alpha\gamma_{ik}$ is the random interaction effect of the ith level of factor A with the kth level of factor C
$\beta\gamma_{jk}$ is the random interaction effect of the jth level of factor B with the kth level of factor C
$\alpha\beta\gamma_{ijk}$ is the random interaction effect of the ith level of factor A with the jth level of factor B and the kth level of factor C
ϵ_{ijkm} is the error associated with the mth replicate of levels i, j, and k of factors A,B, and C respectively

b. The AOV table is shown below:

Source	df	SS	MS	Expected MS	Denom of F
A	4	SSA	SSA/4	$\sigma_\epsilon^2 + 10\sigma_{\alpha\beta\gamma}^2 + 30\sigma_{\alpha\beta}^2 + 60\sigma_{\alpha\gamma}^2 + 180\sigma_\alpha^2$	*
B	5	SSB	SSB/5	$\sigma_\epsilon^2 + 10\sigma_{\alpha\beta\gamma}^2 + 30\sigma_{\alpha\beta}^2 + 50\sigma_{\beta\gamma}^2 + 150\sigma_\beta^2$	*
C	2	SSC	SSC/2	$\sigma_\epsilon^2 + 10\sigma_{\alpha\beta\gamma}^2 + 60\sigma_{\alpha\gamma}^2 + 50\sigma_{\beta\gamma}^2 + 300\theta_C$	*
AB	20	SSAB	SSAB/20	$\sigma_\epsilon^2 + 10\sigma_{\alpha\beta\gamma}^2 + 30\sigma_{\alpha\beta}^2$	MSABC
AC	8	SSAC	SSAC/8	$\sigma_\epsilon^2 + 10\sigma_{\alpha\beta\gamma}^2 + 60\sigma_{\alpha\gamma}^2$	MSABC
BC	10	SSBC	SSBC/10	$\sigma_\epsilon^2 + 10\sigma_{\alpha\beta\gamma}^2 + 50\sigma_{\beta\gamma}^2$	MSABC
ABC	40	SSABC	SSABC/40	$\sigma_\epsilon^2 + 10\sigma_{\alpha\beta\gamma}^2$	MSE
Error	810	SSE	SSE/810	σ_ϵ^2	*
Total	899	SST			

c. The ratios to give the F statistics are as follows:

Source	F
A	*
B	*
C	*
AB	MSAB/MSABC
AC	MSAC/MSABC
BC	MSBC/MSABC
ABC	MSABC/MSE

17.15

a. The model for a 3x6x3 factorial treatment structure with $n = 3$ replications and factor A random and factors B and C fixed is as follows:

$$y_{ijkm} = \mu + \alpha_i + \beta_j + \gamma_k + \alpha\beta_{ij} + \alpha\gamma_{ik} + \beta\gamma_{jk} + \alpha\beta\gamma_{ijk} + \epsilon_{ijkm}; i = 1,2,3; j = 1,\ldots,6; k = 1,2,3; m = 1,2,3$$

Where y_{ijkm} is the response of the mth replication under level i factor A, level j factor B, level k factor C.
μ is the mean of all responses
α_i is the random effect of the ith level of factor A
β_j is the fixed effect of the jth level of factor B
γ_k is the fixed effect of the kth level of factor C
$\alpha\beta_{ij}$ is the random interaction effect of the ith level of factor A with the jth level of factor B
$\alpha\gamma_{ik}$ is the random interaction effect of the ith level of factor A with the kth level of factor C
$\beta\gamma_{jk}$ is the fixed interaction effect of the jth level of factor B with the kth level of factor C
$\alpha\beta\gamma_{ijk}$ is the random interaction effect of the ith level of factor A with the jth level of factor B and the kth level of factor C
ϵ_{ijkm} is the random error associated with the mth replicate of levels $i, j,$ and k of factors A, B, and C respectively

b. The AOV table is shown below:

Source	df	SS	MS	Expected MS	Denom of F
A	2	SSA	SSA/2	$\sigma_\epsilon^2 + 3\sigma_{\alpha\beta\gamma}^2 + 9\sigma_{\alpha\beta}^2 + 18\sigma_{\alpha\gamma}^2 + 54\theta_A$	*
B	5	SSB	SSB/5	$\sigma_\epsilon^2 + 3\sigma_{\alpha\beta\gamma}^2 + 9\sigma_{\alpha\beta}^2 + 9\theta_{BC} + 27\theta_B$	*
C	2	SSC	SSC/2	$\sigma_\epsilon^2 + 3\sigma_{\alpha\beta\gamma}^2 + 18\sigma_{\alpha\gamma}^2 + 9\theta_{BC} + 54\theta_C$	*
AB	10	SSAB	SSAB/10	$\sigma_\epsilon^2 + 3\sigma_{\alpha\beta\gamma}^2 + 9\sigma_{\alpha\beta}^2$	MSABC
AC	4	SSAC	SSAC/4	$\sigma_\epsilon^2 + 3\sigma_{\alpha\beta\gamma}^2 + 18\sigma_{\alpha\gamma}^2$	MSABC
BC	10	SSBC	SSBC/10	$\sigma_\epsilon^2 + 3\sigma_{\alpha\beta\gamma}^2 + 9\theta_{BC}$	MSABC
ABC	40	SSABC	SSABC/40	$\sigma_\epsilon^2 + 3\sigma_{\alpha\beta\gamma}^2$	MSE
Error	88	SSE	SSE/88	σ_ϵ^2	*
Total	161	SST			

c. The ratios to give the F statistics are as follows:

Source	F
A	*
B	*
C	*
AB	MSAB/MSABC
AC	MSAC/MSABC
BC	MSBC/MSABC
ABC	MSABD/MSE

17.17
a. The mixed effects model is more appropriate. Researchers would be concerned about specific chemicals not a population of chemicals. They would want to determine which of the four chemicals is most effective in controlling fire ants.
b. A fixed effects model would be more appropriate if the researcher was only interested in a set of specific locations, such as those with specific environmental conditions, or different levels of human activity or specific soil conditions. The fixed effects model would have both the levels of chemicals and the levels of locations used in the experiment as the only levels of interest. The levels used in the experiment would not be randomly selected from a population of level.

17.19
a. This is a model with day crossed with week. We will avoid the nested model as it is assumed the day effect is the same across weeks.
The model for this situation is:
$y_{ij} = \mu + \alpha_i + \beta_j + \varepsilon_{ijk}; i = 1, \ldots, 12; j = 1, \ldots 5$, where
y_{ij} is the number of non-permit cars on the jth day of week i
μ is the overall mean number of non-permit cars
α_i is the random Week effect, iid $N(0, \sigma_W^2)$ r.v.'s
β_j is the fixed day effect
α_i and β_j are independent

b. The AOV table is shown below

Source	df	SS	MS	EMS	F	p-value
Week	11	287	26.0909	$5\sigma_W^2 + \sigma_\varepsilon^2$	33.28	0.0001
Day	4	79.67	19.9167	$12\theta_D + \sigma_\varepsilon^2$	43.59	0.0001
Error	44	26.33	0.5985	σ_ε^2		
Total	59					

$$H_0: \sigma_W^2 = 0 \text{ vs. } H_A: \sigma_W^2 > 0$$

This test has a p-value of < 0.0001 for both factors which means there is a significant difference in average "suspicious" car appearance over different weeks and across the days of the week.

c. Because the p-value for the hypothesis test run above is significantly less than 0.05, the average number of "suspicious" cars arriving on campus on a weekly basis does not remain constant over the academic year.

d. The only potential violation is the outlier in observation 5.

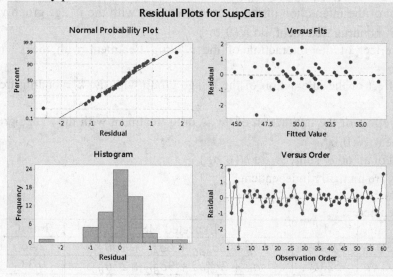

17.21
a. Factor A: Drug Treatment, 3 fixed levels
   Factor B: Age, 2 fixed levels
   Factor C: Nondrug Treatment, 3 random levels
   3 replications of the $t = (3)(3)(2) = 18$ treatments

   y_{ijkm} is the reduction in the systolic blood pressure of the mth patient in the jth age group receiving the ith drug treatment and k^{th} nondrug treatment
   $y_{ijkm} = \mu + \alpha_i + \beta_j + \alpha\beta_{ij} + \gamma_k + \alpha\gamma_{ik} + \beta\gamma_{jk} + \alpha\beta\gamma_{ijk} + \epsilon_{ijkm}$, where

   μ is the overall mean reduction in systolic blood pressure
   α_i is the fixed effect of the ith drug treatment
   β_j is the fixed effect of the jth age group
   $\alpha\beta_{ij}$ is the fixed effect of the interaction of the ith drug treatment with the jth age group γ_k is the random effect of the kth nondrug treatment, iid $N(0, \sigma_C^2)$
   $\alpha\gamma_{ik}$ is the random effect of the interaction of the ith drug treatment with the kth nondrug treatment, $N(0, \sigma_{AC}^2)$
   $\beta\gamma_{jk}$ is the random effect of the interaction of the jth age group with the kth nondrug treatment, $N(0, \sigma_{BC}^2)$
   $\alpha\beta\gamma_{ijk}$ is the random effect of the interaction of the ith drug treatment with the jth age group and the kth nondrug treatment, $N(0, \sigma_{ABC}^2)$
   ϵ_{ijkm} is random effect of all other sources, iid $N(0, \sigma_\epsilon^2)$
   All the random effects are mutually independent.

b. The AOV table is given here:

Source	df	SS	MS	Expected MS	F-test	p-value
Drug	2	431.37	215.69	$\sigma_\epsilon^2 + 3\sigma_{ABC}^2 + 6\sigma_{AC}^2 + 18\theta_A$	2.46	0.201
Age	1	52.02	52.02	$\sigma_\epsilon^2 + 3\sigma_{ABC}^2 + 9\sigma_{BC}^2 + 27\theta_B$	9.33	0.093
Drug*Age	2	77.37	38.69	$\sigma_\epsilon^2 + 3\sigma_{ABC}^2 + 9\theta_{AB}$	3.37	0.139
Nondrug	2	260.48	130.24	$\sigma_\epsilon^2 + 18\sigma_C^2 + 6\sigma_{AC}^2 + 9\sigma_{BC}^2 + 3\sigma_{ABC}^2$	1.60	0.325
Drug*Nondrug	4	350.19	87.55	$\sigma_\epsilon^2 + 6\sigma_{AC}^2 + 3\sigma_{ABC}^2$	7.62	0.037
Age*Nondrug	2	11.15	5.57	$\sigma_\epsilon^2 + 9\sigma_{BC}^2 + 3\sigma_{ABC}^2$	0.48	0.650
Drug*Age*Nondrug	4	45.96	11.49	$\sigma_\epsilon^2 + 3\sigma_{ABC}^2$	3.61	0.014
Error	36	114.67	3.185	σ_ϵ^2		

c. The following Interaction terms are significant:
   Three-way interaction: p-value = 0.014 and Drug*Nondrug: p-value = 0.037
   None of the main effect terms are significant.

d. Because there is a significant interaction between Age, Drug, and Nondrug treatments, which Drug, Nondrug treatment is most effective depends on the age of the patient. The treatment means are given here:

Age	Drug	Nondrug	Mean	Grouping within Age
A1	D1	ND1	34.00	a
A1	D1	ND2	37.00	ab
A1	D1	ND3	40.67	bc
A1	D2	ND1	45.67	c
A1	D2	ND2	47.00	c
A1	D2	ND3	44.00	c
A1	D3	ND1	36.33	a
A1	D3	ND2	44.67	c
A1	D3	ND3	36.00	a

Age	Drug	Nondrug	Mean	Grouping within Age
A2	D1	ND1	35.00	a
A2	D1	ND2	46.33	b
A2	D1	ND3	46.33	b
A2	D2	ND1	47.00	b
A2	D2	ND2	46.00	b
A2	D2	ND3	43.67	b
A2	D3	ND1	36.67	a
A2	D3	ND2	45.67	b
A2	D3	ND3	36.33	a

e. Recommending a treatment based on this data is difficult due to the presence of the 3-factor interaction. Different recommendations would be made based on the age group and the age group is so large, it may be beneficial to further discretize the ages to make better recommendations. Based on the results above, combinations in group 'c' for A1 and group 'b' for A2 seem to yield the largest mean reduction.

17.23
a. A test for the equality of the treatment means in the fixed-effects model is
$$H_0: \alpha_1 = \cdots = \alpha_t = 0 \text{ vs. } H_a: \text{at least one } \alpha_i \neq 0$$
In the fixed effects model, we are testing the difference in the means for the t treatments used in the experiment.

b. A test concerning the variability in the population of means in the random effects model is:
$$H_0: \sigma_A^2 = 0 \text{ vs. } H_a: \sigma_A^2 > 0$$
In the random effects model, we are testing the difference in a population of means from which the t treatments used in the experiment were randomly selected.

17.25

a. This is a completely randomized mixed model with 2 replicates per treatment.
   Factor A: Temperature is fixed with 5 levels
   Factor B: Pane Design is random with 5 levels
   The AOV table is given here:

Source	df	SS	MS	EMS	F	P
Temp	4	39.7788	9.9447	$\sigma_\epsilon^2 + 2\sigma_{AB}^2 + 10\theta_A$	14.50	0.0001
Panes	4	7.3228	1.8307	$\sigma_\epsilon^2 + 10\sigma_B^2 + 2\sigma_{AB}^2$	2.97	0.052
Interaction	16	10.9712	0.6857	$\sigma_\epsilon^2 + 2\sigma_{AB}^2$	2.97	0.0072
Error	25	5.7800	0.2312	σ_ϵ^2		
Total	49	63.8528				

b. The interaction between Temperature and Pane Design is significant (p-value = 0.0072). Only the Temperature main effect is significant: Temperature (p-value = 0.0001). The main effect for Pane Design is not significant (p-value = 0.052).

c. In Exercise 14.31, all three terms were significant. The difference is that in this case the inferences are concerning the population of pane designs and not just the five designs used in the study.

d. If there are a large number of commercial thermal pane designs available, then it would be reasonable to randomly select a few for comparison in the study. If the only pane designs available are the five used in the study, then the fixed effects model would be the appropriate model.

17.27

a. This is 2 reps of a completely randomized mixed model with
   Factor A: Operator is random effect with 3 levels
   Factor B: Machine is random effect with 4 levels
   The model is $y_{ijk} = \mu + \alpha_i + \beta_j + \alpha\beta_{ij} + \varepsilon_{ijk}$
   y_{ijk} is the strength of the kth solder of the ith operator using the jth machine
   μ is the mean strength over all operators and machines
   α_i is the random effect of the ith operator, iid $N(0, \sigma_A^2)$ r.v.'s
   β_j is the random effect of the jth machine, iid $N(0, \sigma_B^2)$ r.v.'s
   $\alpha\beta_{ij}$ is the random interaction effect of the ith operator with the jth machine iid $N(0, \sigma_{AB}^2)$ r.v.'s with $\sum_i \alpha\beta_{ij} = 0$
   ϵ_{ijk} is the random effect of all other factors on strength
   $\alpha_i's$, $\beta_j's$, $\alpha\beta_{ij}$'s and ϵ_{ijk}'s are all independent

b. The AOV table is given here:

Source	df	SS	MS	EMS	F	P
Operator	2	160.333	80.167	$\sigma_\epsilon^2 + 2\sigma_{AB}^2 + 8\sigma_A^2$	10.77	0.0103
Machine	3	12.458	4.153	$\sigma_\epsilon^2 + 2\sigma_{AB}^2 + 6\sigma_B^2$	0.56	0.6607
Interaction	6	44.667	7.444	$\sigma_\epsilon^2 + 2\sigma_{AB}^2$	1.96	0.1513
Error	12	45.500	3.792	σ_ϵ^2		
Total	23	262.958				

c. The interaction between Operator and Machine is not significant (p-value = 0.1513), the main effect of Operator is significant (p-value = 0.0103), but the main effect of Machine is not significant (p-value = 0.6607). There is a significant difference in the mean strength of the solder between the operators with these differences consistent across different machines. There is not significant evidence that the mean strengths differ between different machines.

17.29
  a. Nested design with Analyses nested within Location. The model is
  $y_{ij} = \mu + \alpha_i + \beta_{j(i)}$, where
  y_{ij} is the sulfur content of the jth analysis at location i
  μ is the mean sulfur content over all analyses and locations
  α_i is the effect of the ith location
  $\beta_{j(i)}$ is the effect of the jth analysis at location i

  b. The AOV table is given here:

Source	df	SS
Location	6 − 1 = 5	SSL
Analysis with Location	6(4 − 1) = 18	SSA(L)
Total	24 − 1 = 23	SST

17.31
  a. This is a nested design with Samples nested within Batches.
  b. The model for this situation is:
  $y_{ijk} = \mu + \alpha_i + \beta_{j(i)} + \epsilon_{ijk}$, where
  y_{ijk} is the hardness of the kth tablet from Sample j selected from Batch i
  μ is the overall mean hardness
  α_i is the random Batch effect, iid $N(0, \sigma_B^2)$ r.v.'s
  $\beta_{j(i)}$ is the random Sample within Batch effect, iid $N(0, \sigma_S^2(B))$ r.v.'s
  ϵ_{ijk} is the random effect due to all other factors, iid $N(0, \sigma_\epsilon^2)$ r.v.'s
  $\alpha_i, \beta_{j(i)},$ and ϵ_{ijk} are all independent

  c. The AOV table is given here:

Source	df	SS	MS	F	p-value
Batch	2	9095.5238	4547.7619	101.635	0.0001
Sample	6	268.4762	44.7460	1.533	0.1851
Error	54	1576.00	29.1852		
Total	62	10940.00			

  d. There is significant evidence (p-value < 0.0001) that the batches produced different mean hardness values. There does not appear to be a significant (p-value = 0.1851) variation in the samples within the batches.

  The variance components are given here:

Source	Var Component	% of Total
Batch	214.429	87.22
Sample	2.223	0.90
Error	29.185	11.87
Total	245.837	

17.33

a. Fixed: Soil Type(So)—Because the specific soil types are selected by the researcher deterministically
Random: Site (Si)—Because the sites are randomly chosen from a given soil type

b. The factors are nested (Site within Soil). This is nested because only certain sites contain a given soil type.

c. The AOV table is given below:

Source	df	Expected MS
So	2	$\sigma_\epsilon^2 + 5\sigma_{So(Si)}^2 + 25\theta_{So}$
Si(So)	12	$\sigma_\epsilon^2 + 5\sigma_{So(Si)}^2$
Error	60	σ_ϵ^2
Total	74	

17.35

a. The model for a 5x3x8 factorial treatment structure with $n = 1$ replications and factors A random and B and C fixed is as follows:

$$y_{ijk} = \mu + \alpha_i + \beta_j + \gamma_k + \alpha\beta_{ij} + \alpha\gamma_{ik} + \beta\gamma_{jk} + \epsilon_{ijkm}; i = 1,\ldots,5; j = 1,2,3; k = 1,\ldots,8$$

Where y_{ijk} is the response of the ith dentist, j^{th} method, k^{th} alloy.

The 3 factor interaction was not estimated due to only 1 replicate per combination.

μ is the mean of all responses
α_i is the random effect of the ith dentist
β_j is the fixed effect of the jth method
γ_k is the fixed effect of the kth alloy
$\alpha\beta_{ij}$ is the random interaction effect of the ith dentist with the jth method
$\alpha\gamma_{ik}$ is the random interaction effect of the ith dentist with the kth alloy
$\beta\gamma_{jk}$ is the random interaction effect of the jth method with the kth alloy
ϵ_{ijk} is the random error associated with the observation with levels $i, j,$ and k of factors A, B, and C respectively

b. The AOV table is shown below.
Analysis of Variance

```
Source           DF   Adj SS   Adj MS   F-Value   P-Value
  Dentist         4   217576    54394     1.79     0.240   x
  Method          2   597615   298808     9.07     0.009
  Alloy           7   220338    31477     4.22     0.003
  Dentist*Method  8   263441    32930     3.30     0.004
  Dentist*Alloy  28   208814     7458     0.75     0.797
  Method*Alloy   14   209773    14984     1.50     0.140
Error            56   558258     9969
Total           119  2275815

x Not an exact F-test.
```

Source	EMS
Dentist	$\sigma_\epsilon^2 + 8\sigma_{\alpha\beta}^2 + 3\sigma_{\alpha\gamma}^2 + 24\sigma_\alpha^2$
Method	$\sigma_\epsilon^2 + 8\sigma_{\alpha\beta}^2 + 5\theta_{BC} + 40\theta_B$
Alloy	$\sigma_\epsilon^2 + 3\sigma_{\alpha\gamma}^2 + 5\theta_{BC} + 15\theta_C$
D*M	$\sigma_\epsilon^2 + 8\sigma_{\alpha\beta}^2$
D*A	$\sigma_\epsilon^2 + 3\sigma_{\alpha\gamma}^2$
M*A	$\sigma_\epsilon^2 + 5\theta_{BC}$
Error	σ_ϵ^2

c. The p-value for the method-alloy interaction is 0.14 > 0.05. Therefore, there is not significant evidence of an interaction between method and alloy.

17.37
a. Lab is nested within Methods. Source is crossed with lab and method.
$y_{ijkl} = \mu + \alpha_i + \beta_j + \gamma_{k(j)} + \alpha\beta_{ij} + \alpha\gamma_{ik(j)} + \epsilon_{ijkl}; i = 1,2,3; j = 1,2,3,4; k = 1,...,20; l = 1,2$
Where y_{ijkl} is the ecoli measurement of the ith source, j^{th} method, k^{th} lab nested within method, l^{th} replicate.
μ is the mean of all responses
α_i is the fixed effect of the ith source
β_j is the fixed effect of the jth method
γ_k is the random effect of the kth lab
$\alpha\beta_{ij}$ is the fixed interaction effect of the ith source with the jth method
$\alpha\gamma_{ik(j)}$ is the random interaction effect of the ith source with the kth lab
ϵ_{ijkl} is the random error associated with the observation with levels i, j, and k of factors A,B, and C respectively

b. The conditions for testing hypotheses and constructing confidence intervals appear satisfied for the nested model.

c. The AOV table is shown below.
Analysis of Variance

```
Source                DF   Adj SS   Adj MS   F-Value   P-Value
  Source               2    9.456   4.7281    12.05    0.000
  Method_1             3   11.652   3.8839     8.14    0.002
  Source*Method_1      6   27.219   4.5365    11.56    0.000
  Lab(Method_1)       16    7.631   0.4770     1.22    0.309
  Source*Lab(Method_1) 32  12.555   0.3923     1.15    0.318
Error                 60   20.535   0.3423
Total                119   89.048
```

Model Summary

```
       S    R-sq   R-sq(adj)  R-sq(pred)
0.585021  76.94%      54.26%       7.76%
```

Source	EMS
Source	$\sigma_\epsilon^2 + 10\sigma_{\alpha\beta}^2 + 2\sigma_{\alpha\gamma(j)}^2 + 40\theta_S$
Method	$\sigma_\epsilon^2 + 10\sigma_{\alpha\beta}^2 + 30\theta_M$
Lab	$\sigma_\epsilon^2 + 2\sigma_{\alpha\gamma(j)}^2 + 6\sigma_\gamma^2$
Source*Method	$\sigma_\epsilon^2 + 10\sigma_{\alpha\beta}^2$
Source*Lab	$\sigma_\epsilon^2 + 2\sigma_{\alpha\gamma(j)}^2$
Error	σ_ϵ^2

d. The interaction of source and method is significant. The lab is not significant in determining Ecoli and neither are the interactions.

e. Overall, the assessment methods are different, and their effect on the response changes with different sources.

Chapter 18

Split-Plot, Repeated Measures, and Crossover Designs

18.1
  a. First, assign the numbers 1-40 to the 40 wholeplots. Create a random permutation of the numbers 1-40 and for the first 10 numbers (plots) in the random permutation, assign wholeplot treatment A_1, for the second 10 numbers (plots), assign wholeplot treatment A_2, for the third 10 numbers (plots), assign wholeplot treatment A_3, for the last 10 numbers (plots), assign wholeplot treatment A_4. Within each wholeplot, label the experimental units 1,2,3. Create a random permutation of these numbers to assign the B subplot treatments.
  b. The model is as follows:
  $$y_{ijk} = \mu + \tau_i + \delta_{ik} + \gamma_j + \tau\gamma_{ij} + \epsilon_{ijk}; i = 1,2,3,4; j = 1,2,3; k = 1,\ldots,10$$
  τ_i is the fixed effect of the ith level of A
  γ_j is the fixed effect of the jth level of B
  $\tau\gamma_{ij}$ is the fixed interaction effect of the ith level of A on the jth level of B
  δ_{ik} is the random effect for the kth wholeplot receiving the ith level of A. The δ_{ik} are independent normal with mean 0 and variance σ_δ^2
  ϵ_{ijk} is the random error effect. They are independent normals with mean 0 and variance σ_ϵ^2.
  c. The AOV table is shown below:

Source	SS	df	EMS
A	SSA	3	$\sigma_\epsilon^2 + 3\sigma_\delta^2 + 30\theta_\tau$
Wholeplot Error	SS(A)	36	$\sigma_\epsilon^2 + 3\sigma_\delta^2$
B	SSB	2	$\sigma_\epsilon^2 + 40\theta_\gamma$
AB	SSAB	6	$\sigma_\epsilon^2 + 10\theta_{\tau\gamma}$
Subplot Error	SSE	72	σ_ϵ^2
Total	SSTot	119	

18.3
  a. For each of the three blocks, assign each experimental unit (wholeplot) a number 1, 2, 3, 4. Obtain a random permutation of the numbers 1, 2, 3, 4 and assign treatment A_1 to the first, A_2 to the second, etc. Once the wholeplots have been randomized in the blocks, assign each subplot a number 1, …, 6. Obtain a random permutation of those numbers 1,…, 6 and assign treatment B_1 to the first two, B_2 to the second two, and so on.
  b. The model is as follows:
  $$y_{ijk} = \mu + \tau_i + \beta_j + \tau\beta_{ij} + \gamma_k + \tau\gamma_{ik} + \epsilon_{ijk}; i = 1,2,3,4 j = 1,2,3; k = 1,2,3,$$
  τ_i is the fixed effect of the ith level of A
  β_j is the block effect of the jth block
  $\tau\beta_{ij}$ is the interaction between the ith level of A and the jth block effect
  γ_k is the fixed effect of the kth level of B
  $\tau\gamma_{ik}$ is the fixed interaction effect of the ith level of A on the kth level of B

c. The AOV table is shown below:

Source	SS	df	EMS
Block	SSBlock	2	$\sigma_\epsilon^2 + 12\sigma_{\tau\beta}^2$
A	SSA	3	$\sigma_\epsilon^2 + 3\sigma_{\tau\beta}^2 + 9\theta_\tau$
A*Block	SSA*Block	6	$\sigma_\epsilon^2 + 3\sigma_{\tau\beta}^2$
B	SSB	2	$\sigma_\epsilon^2 + 12\theta_\gamma$
AB	SSAB	6	$\sigma_\epsilon^2 + 3\theta_{\tau\gamma}$
Error	SSE	52	σ_ϵ^2
Total	SSTot	71	

18.5
a. The SAS Output is provided below
The GLM Procedure

Dependent Variable: y

Source	DF	Sum of Squares	Mean Square	F Value	Pr > F
Model	20	52.25750000	2.61287500	47.97	<.0001
Error	27	1.47062500	0.05446759		
Corrected Total	47	53.72812500			

R-Square	Coeff Var	Root MSE	y Mean
0.972628	5.565018	0.233383	4.193750

Source	DF	Type III SS	Mean Square	F Value	Pr > F
RATION	2	8.79875000	4.39937500	80.77	<.0001
RATION*REP	9	38.26187500	4.25131944	78.05	<.0001
AGE	3	4.47229167	1.49076389	27.37	<.0001
AGE*RATION	6	0.72458333	0.12076389	2.22	0.0722

Split-Plot Design   22

The GLM Procedure

Source	Type III Expected Mean Square
RATION	Var(Error) + 4 Var(RATION*REP) + Q(RATION,AGE*RATION)
RATION*REP	Var(Error) + 4 Var(RATION*REP)
AGE	Var(Error) + Q(AGE,AGE*RATION)
AGE*RATION	Var(Error) + Q(AGE*RATION)

The GLM Procedure
Tests of Hypotheses for Mixed Model Analysis of Variance

Dependent Variable: y

```
    Source              DF      Type III SS     Mean Square     F Value     Pr > F

*   RATION              2        8.798750        4.399375        1.03       0.3940

    Error               9       38.261875        4.251319
Error: MS(RATION*REP)
* This test assumes one or more other fixed effects are zero.

    Source              DF      Type III SS     Mean Square     F Value     Pr > F

    RATION*REP          9       38.261875        4.251319       78.05       <.0001
*   AGE                 3        4.472292        1.490764       27.37       <.0001
    AGE*RATION          6        0.724583        0.120764        2.22       0.0722

    Error: MS(Error)   27        1.470625        0.054468
* This test assumes one or more other fixed effects are zero.
```

b. The p-value to test for the significance of the interaction is 0.0722 which implies the interaction between ration and age is not significant. This will allow inference on the main effects.

c. The p-value to test the significance of the effect of rations is 0.3940 > 0.05. Therefore, it can be concluded that there is not a significant effect of ration on mean shear force.

d. The p-value to test the significance of the effect of age is < 0.0001. Therefore, it can be concluded that there is a significant effect of age on mean shear force.

18.7
a. A profile plot of the water loss data is given here:

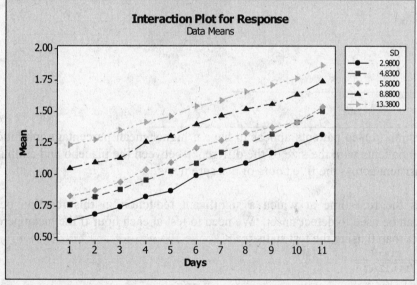

b. There appears to be an increase in the mean water loss as the level of saturation deficit increases.

18.9
a. The mean and standard deviation of percentage inhibition by Treatment and Time are given here:

Treatment (Means)	Time				
	1	2	3	4	8
Antihistamine	20.70	28.57	31.24	29.44	25.63
Placebo	-0.76	12.55	18.23	24.79	17.57
Treatment (St. Dev.)	1	2	3	4	8
Antihistamine	23.98	12.00	14.30	12.65	14.26
Placebo	12.26	10.43	10.83	6.91	7.83

The antihistamine treated patients uniformly, across all five hours, have larger mean percentage inhibition than the placebo treated patients. The pattern for the standard deviations is similar with somewhat higher values during the first hour after treatment.

b. A profile plot of the skin sensitivity data is given here:

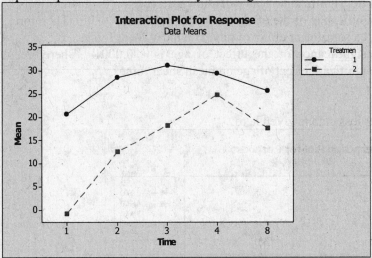

c. Yes, the antihistamine treated patients appear to have a higher mean percentage inhibition than the placebo treated patients with the size of the difference between the placebo and antihistamine patients fairly consistent across the five hours of measurements.

18.11 Defining onset as the first time at which a significant reduction in this reaction occurs, a Bonferroni t-test can be used to detect onset. We need to test at each hour if the mean percentage inhibition is greater than 0 using the test statistic:

$$t = \frac{\bar{y}_{i,k} - 0}{\sqrt{MSE}/\sqrt{n_{ik}}} = \frac{\bar{y}_{i,k} - 0}{\sqrt{141.12}/\sqrt{10}}$$

Since we are making 10 tests, we will use $\alpha = \frac{0.05}{10} = 0.005$ and critical value $t_{0.005,72} = 2.646$. We will declare that the mean percentage inhibition is greater than 0 whenever $t > 2.646$. The results are summarized in the following table:

	Time				
	1	2	3	4	8
Antihistamine (Mean)	20.70	28.57	31.24	29.44	25.63
t-value	5.51	7.61	8.32	7.84	6.82
Significant	Yes	Yes	Yes	Yes	Yes
Placebo (Mean)	-0.76	12.55	18.23	24.79	17.57
t-value	-0.20	3.34	4.85	6.60	4.68
Significant	No	Yes	Yes	Yes	Yes

Onset occurs during the first hour for the antihistamine treatment whereas onset occurs at the second hour for the placebo. The antihistamine has an earlier onset and a consistently larger mean percentage inhibition than the placebo at each hour.

18.13
   a. This is a two period crossover design with two treatments, two treatment sequences, and eight patients per sequence.

   b. A model for this experiment is given here:
   $y_{ijk} = \mu + \alpha_i + \beta_j + \gamma_k + \pi_{m(k)} + \epsilon_{ijkm}$, where
   y_{ijkm} is the sleep duration of the mth patient during period k under the ith treatment in sequence k
   α_i is the fixed effect of the ith treatment
   β_j is the fixed effect of Period j
   γ_k is the fixed effect of the kth sequence
   $\pi_{m(k)}$ is the random effect of patient m in k^{th} Sequence, iid $N(0, \sigma_P^2)$ r.v.'s
   β_k is the fixed effect of Hour k
   ϵ_{ijkm} is the random effect of all other factors on sleep duration, iid $N(0, \sigma_\epsilon^2)$ r.v.'s

   c. This design assumes that there is no interaction between treatment and period.

18.15
   a. Paired t-test could be used.
   b. Test $H_0: \mu_d \leq 0$ versus $H_a: \mu_d > 0$ with test statistic:
   $t = \frac{\bar{d} - 0}{s_d/\sqrt{16}} = \frac{0.43125}{0.32806/\sqrt{16}} = 27.648$ with df = 16 − 1 = 15.
   p-value = $\Pr(t_{15} \geq 27.648) < 0.0001$. There is significant evidence that the treatment has a larger mean sleep duration than the placebo.

18.17 In period one, patients in sequence one received the drug and those in sequence two received the placebo. A two-sample t-test could be conducted using just the period one data:
$s_1^2 = 10(0.3594)^2 = 1.2917, s_2^2 = 10(0.1955)^2 = 0.3822,$

$$s_p = \sqrt{\frac{7(1.2917) + 7(0.3822)}{14}} = 0.9148, t = \frac{7.3250 - 7.150}{0.9148\sqrt{\frac{1}{8} + \frac{1}{8}}} = 0.428 \text{ with}$$

df = 14 $\Rightarrow$ p-value = $\Pr(t_{14} \geq 0.428) = 0.3376 \Rightarrow$ There is not significant evidence of a difference in mean sleep duration between the drug treated patients and the placebo treated patients.

18.19 The analysis of variance table is given here:

Source	DF	Adj SS	Adj MS	F	P
Sequence	2	7246.8	3623.4	3.33	0.071
Patient (Sequence)	12	13072.2	1089.4	*	*
Formulation	2	41.7	20.8	0.31	0.733
Period	2	12110.3	6055.1	91.21	0.000
Error	26	1726.0	66.4		
Total	44	34197.0			

Based on the results in the AOV table, the conclusions based on the profile plot are confirmed. There is a significant Period effect (p-value < 0.0001), the effect due to Formulations is not significant (p-value = 0.733), and there is not an effect due to Sequence (p-value = 0.071).

18.21
a. The plot of formulation means by period for each sequence is shown below:

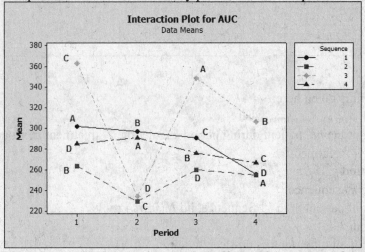

b. There does not appear to be a period effect. The average of the four treatment means within each period yield similar values with the exception of the large values for C in period 1 and A in period 2.

c. The means and standard deviations by treatment are as follows:

Descriptive Statistics: AUC

```
Variable     Treatmen    Mean    StDev
AUC          A           297.0   129.0
             B           284.3   140.4
             C           284.2   135.0
             D           259.5   115.5
```

The formulations do not appear to have significantly different means as the values are reasonably close when observing the large standard deviations. Treatment D appears lower than the others but the large standard deviation leads one to conclude no difference.

18.23 The output from SAS is shown below:

Source	DF	Sum of Squares	Mean Square	F Value	Pr > F
Model	34	1274442.436	37483.601	5.85	<.0001
Error	69	442326.368	6410.527		
Corrected Total	103	1716768.804			

R-Square	Coeff Var	Root MSE	AUC Mean
0.742349	28.46685	80.06577	281.2597

Source	DF	Type III SS	Mean Square	F Value	Pr > F
Sequence	3	49286.962	16428.987	2.56	0.0618
Subject(Sequence)	22	1162481.582	52840.072	8.24	<.0001
Treatment	3	19187.195	6395.732	1.00	0.3992
Period	3	28149.496	9383.165	1.46	0.2320
carryover	3	15337.201	5112.400	0.80	0.4995

The p-value for the Carryover effect is 0.4995 > 0.05 which implies there is no significant evidence of a carryover effect. Therefore, it can be assumed the drugs of a previous treatment have washed out of the system prior to the next treatment being administered.

18.25
a. Yes, the distribution of age, gender, and duration of illness are similar for the Drug and Placebo groups.
b. For the age data:
$\bar{y}_1 = 37.2, s_1 = 10.5, n_1 = 15$
$\bar{y}_2 = 39.50, s_2 = 9.6, n_2 = 15 \Rightarrow$

$$s_P = \sqrt{\frac{14(10.5)^2 + 14(9.6)^2}{28}} = 10.06 \Rightarrow$$

$$t = \frac{37.2 - 39.5}{10.06\sqrt{\frac{1}{15} + \frac{1}{15}}} = -0.63 \Rightarrow p-value = 2\Pr(t_{28} \geq 0.63) = 0.2669 \Rightarrow$$

No significant difference in the means for age between the Drug and Placebo groups.

For the duration data:
$\bar{y}_1 = 10.7, s_1 = 6.5, n_1 = 15$
$\bar{y}_2 = 11.5, s_2 = 7.3, n_2 = 15 \Rightarrow$

$$s_P = \sqrt{\frac{14(6.5)^2 + 14(7.3)^2}{28}} = 6.911 \Rightarrow$$

$$t = \frac{10.7 - 11.5}{6.911\sqrt{\frac{1}{15} + \frac{1}{15}}} = -0.32 \Rightarrow p-value = 2\Pr(t_{28} \geq 0.32) = 0.3757 \Rightarrow$$

No significant difference in the means for duration between the Drug and Placebo groups.

c. Compare the proportion of males and females in the two groups using a 2x2 contingency table:

	Males	Females
Drug	20	10
Placebo	16	14

The Chi-square test of homogeneity of proportions yields $\chi^2 = 1.11$ with df = $(2-1)(2-1)$ and p-value = $\Pr(\chi_1^2 \geq 1.11) = 0.292$. Thus, there is not significant evidence that the frequency of Males and Females is different for the Drug and Placebo groups.

18.27

a. We want to test $H_0: \mu_1 \leq \mu_2$ versus $H_a: \mu_1 > \mu_2$, where μ_1 and μ_2 are the mean differences between the number of seizures baseline and number of seizures at five months for the Drug and Placebo groups, respectively.

Drug Group: $D_{1i} = B_{1i} - X_{1i} \Rightarrow \bar{D}_1 = 13.67 \quad s_1 = 6.83 \quad n_1 = 15$
Placebo Group: $D_{2i} = B_{2i} - X_{12i} \Rightarrow \bar{D}_2 = 5.80 \quad s_2 = 4.16 \quad n_2 = 15$

$$s_P = \sqrt{\frac{14(6.83)^2 + 14(4.16)^2}{28}} = 5.655 \Rightarrow$$

$$t = \frac{13.67 - 5.80}{5.655\sqrt{\frac{1}{15} + \frac{1}{15}}} = 3.82 \Rightarrow p-value = \Pr(t_{28} \geq 3.82) = 0.0003 < .01 = \alpha \Rightarrow$$

There is significant evidence that the drug has a larger reduction in the mean number of seizures from baseline to month 5 than the Placebo group. Thus, we have a somewhat different conclusion since the p-value is now sufficiently small enough to support the claim that the drug reduces number of seizures.

b. The comparison of the maximums for the two groups would not be a good idea since our conclusions about the effectiveness of the drug would be based on the most extreme observation in the study. An improved approach which would take into account the variability in seizure rates would be to compare the median change in seizure rates from baseline to month five of the two groups using the Wilcoxon Rank Sum test. Using medians would eliminate the effect of having a few very large or small rates on the mean but would still provide a more valid comparison of the two groups than would the maximum. The median reduction from baseline to month 5 for the Drug Group is 13 and 4 for the Placebo Group. A 95% C.I. on the difference in the two medians is (4,12) and the Wilcoxon Rank Sum test has a p-value = 0.0003 for testing the research hypothesis that the median reduction in the number of seizures from baseline to month 5 for the Drug Group is greater than for the Placebo Group. Thus, we have very strong evidence that the Drug is effective in reducing the number of seizures. In fact, we have 95% confidence that the median reduction is between 4 and 12 seizures larger than the median reduction found in the Placebo Group.

18.29
a. This is a randomized block split-plot design. Tasters are the blocks, Fat levels are the whole plot factor with experimental unit a portion of meat, and Method of cooking is the split-plot factor with experimental unit 1/3 of a portion of meat. There is a single replication of the experiment.

b. $y_{ijk} = \mu + \alpha_i + \beta_j + \alpha\beta_{ij} + \gamma_k + \gamma\alpha_{ik} + \epsilon_{ijk}$, where
y_{ijk} is the taste score from the kth tester for a meat sample having the ith fat level cooked using method j
α_i is the fixed effect of the ith fat level
β_j is the fixed effect of the jth cooking method
$\alpha\beta_{ij}$ is the interaction effect of the ith fat level with j^{th} cooking method
γ_k is the fixed effect of the kth taster
$\gamma\alpha_{ik}$ is the whole plot random effect
ϵ_{ijk} is the random effect due to all other factors

c. Note that the computer output has the wrong F-test for the main effects due to Fat level. The correct F-test has MSF/MST*F as given here:

Source	DF	SS	MS	F	P
Taster (T)	3	25.6389	8.5462	*	*
Fat (F)	2	146.000	73.000	0.97	0.4317
T*F	6	451.778	75.296		
Method (M)	2	22.167	11.083	1.10	0.3551
F*M	4	3.333	0.8333	0.08	0.9868
Error	18	165.333	9.185		
Total	35	1097.000			

d. The interaction between Method of Cooking and Level of Fat is not significant (p-value = 0.9868). The main effects of Method of Cooking and Level of Fat are both non-significant (p-value = 0.3551, p-value = 0.4317, respectively). Thus, there is not significant evidence that either Method of Cooking or Level of Fat have an effect on the taste of the meat.

18.31
a. The SAS output for the repeated measures ANOVA is shown below.
The GLM Procedure

Dependent Variable: Y % LDHLEAKAGE

Source	DF	Sum of Squares	Mean Square	F Value	Pr > F
Model	159	13.94953482	0.08773292	23.41	<.0001
Error	288	1.07944286	0.00374807		
Corrected Total	447	15.02897768			

R-Square	Coeff Var	Root MSE	Y Mean
0.928176	24.82100	0.061221	0.246652

Source	DF	Type III SS	Mean Square	F Value	Pr > F
REP	3	0.90033661	0.30011220	80.07	<.0001
CCL4	3	0.98850804	0.32950268	29.49	<.0001
CHCL3	3	0.87396875	0.29132292	26.08	<.0001
CCL4*CHCL3	9	1.87392946	0.20821438	18.64	<.0001
REP(CCL4*CHCL3)	45	0.50272054	0.01117157	2.98	<.0001
TIME	6	6.44726830	1.07454472	286.69	<.0001
CCL4*TIME	18	0.54938884	0.03052160	8.14	<.0001
CHCL3*TIME	18	0.97796562	0.05433142	14.50	<.0001
CCL4*CHCL3*TIME	54	0.83544866	0.01547127	4.13	<.0001

Note: The denominator for the F statistics corresponding to the main effects and the interaction of CCl4 and ChCl3 comes from the MSRep(CCl4*ChCl3) term.

The GLM Procedure

Source	Type III Expected Mean Square
REP	Var(Error) + 7 Var(REP(CCL4*CHCL3)) + 112 Var(REP)
CCL4	Var(Error) + 7 Var(REP(CCL4*CHCL3)) + Q(CCL4,CCL4*CHCL3,CCL4*TIME,CCL4*CHCL3*TIME)
CHCL3	Var(Error) + 7 Var(REP(CCL4*CHCL3)) + Q(CHCL3,CCL4*CHCL3,CHCL3*TIME,CCL4*CHCL3*TIME)
CCL4*CHCL3	Var(Error) + 7 Var(REP(CCL4*CHCL3)) + Q(CCL4*CHCL3,CCL4*CHCL3*TIME)
REP(CCL4*CHCL3)	Var(Error) + 7 Var(REP(CCL4*CHCL3))
TIME	Var(Error) + Q(TIME,CCL4*TIME,CHCL3*TIME,CCL4*CHCL3*TIME)
CCL4*TIME	Var(Error) + Q(CCL4*TIME,CCL4*CHCL3*TIME)
CHCL3*TIME	Var(Error) + Q(CHCL3*TIME,CCL4*CHCL3*TIME)
CCL4*CHCL3*TIME	Var(Error) + Q(CCL4*CHCL3*TIME)

The p-value for the 3-way interaction between time, CCl4, and ChCl3 is < 0.0001 which implies it is significant. Therefore, we can conclude not only that the time effect is significant but also that the interaction between the chemicals change over time. Because the interaction between the chemicals is significant, there is no reason to interpret the p-values of the main effects individually.

Chapter 18: Split-Plot, Repeated Measures, and Crossover Designs   245

b. The conditions of equal variance and normality of residuals appear to be satisfied as shown below in the following SAS diagnostic plots.

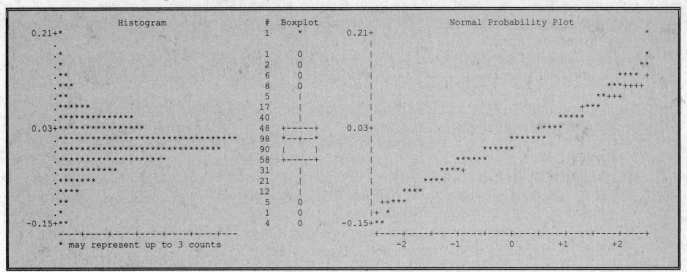

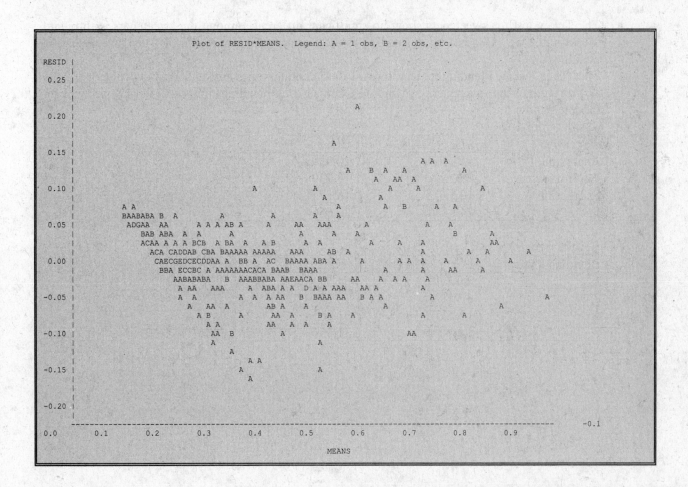